KEEPING HEALTHY
HONEY BEES

**COMMUNITIES, CULTURAL SERVICES
and ADULT EDUCATION**

**This book should be returned on or before
the last date shown below.**

NA2

MARKET RASEN
9/10

2 6 NOV 2010

- 6 SEP 2013

2 1 MAR 2014

29/12/20
22/2/21

WITHDRAWN
FOR SALE

To renew or order library books please telephone 01522 782010
or visit www.lincolnshire.gov.uk
You will require a Personal Identification Number.
Ask any member of staff for this.

Northern Bee B(

D1334222

04642961

Published in the United Kingdom by
Northern Bee Books, Scout Bottom Farm,
Mytholmroyd, West Yorkshire HX7 5JS
Tel: 01422 882751   Fax: 01422 886157
www.GroovyCart.co.uk/beebooks

ISBN - 978-1-904846-54-3
Keeping Healthy Honey Bees,
by Dr. David Aston and Dr. Sally Bucknall

First Published 2010

set in 12pt Bembo

PHOTOGRAPH ACKNOWLEDGMENTS

Photographs courtesy of John Phipps and David Aston except for
plates 4, 7, 8a & b, 16, 17 and 23
(Crown copyright Defra, National Bee Unit in various publications);
Plate 18 Death's-head Hawk-moth (David Chesmore);
Plates 24 and 25 Small Hive Beetle (Kim Flottum);
page 9 Training beekeepers (Board of the National Diploma in Beekeeping)
page 13 Worker bee and page 101 Useful equipment
(The British Beekeepers' Association); and
page 142 Catching swarms (Left) (Sally Bucknall).

Printed by Charlesworth Group, Wakefield (UK)

KEEPING HEALTHY HONEY BEES

David Aston

Sally Bucknall

2010

Northern Bee Books

ACKNOWLEDGMENTS

The writing and publication of all books requires an input from many people as well as the authors and this book would not have been possible without such contributions.

In particular we wish to thank Jeremy Burbidge of Northern Bee Books for supporting its publication and David Miller for his advice and design input.

The majority of the photographs were kindly contributed by John Phipps and for this gesture we thank him very warmly.

Wressle
East Riding of Yorkshire
February 2010

KEEPING HEALTHY HONEY BEES
CONTENTS AND INDEX

SECTION 2

APPROACHES TO
BEE HEALTH MANAGEMENT   33

SECTION 3

SECTION 5

SECTION 7

SECTION 8

SECTION 9

LIST OF TABLES

PROLOGUE

The life of a worker honey bee is short and exhausting. A summer bee born at the start of the forage season will spend the last quarter of her life collecting pollen and nectar; a life that can last a mere 4 – 5 weeks, after which she will die of exhaustion. Over-wintering bees, born from August onwards, live longer. It is they who will rear the brood from eggs which may be laid by the queen as early as January of the following year. The role these bees play is vital to the survival of the colony as long as they are able to fulfil their individual potential; they die but they ensure that the colony lives on. Apart from genetic flaws, especially in the queen, anything that changes this rhythm is a reaction to a stimulus from outside the colony, it can be the result of environmental factors, the impact of disease or the activities of humans, and sometimes the inexperience of beekeepers.

Environmental influences include intermittently cold and warm periods in winter during which vital stores may be used up and colonies may starve, and cold wet summers when insufficient stores are built up, colonies are weakened and may either fall victim to disease or to starvation. Yet good summer weather is not always the key to colony survival; without access to high protein forage throughout the active season colonies will not prosper. The loss of wild flowers due to intensification of agriculture, the extensive growing of monocultures and bad hedgerow and field margin management has impacted heavily on honey bee forage in recent years. Colony losses will continue to occur without the active input of farmers, land managers and gardeners willing to grow a wider range of crops and plants, to manage hedgerows sympathetically to provide suitable bee forage and to support biodiversity in general.

Recently there has been much speculation as to the effect on bees of certain chemicals, such as pesticides, introduced into the environment, and responsible farmers will notify beekeepers before using them. However the effects on bees of some current pesticides are not yet fully understood at the time of writing, and further research into their sub-lethal effects is needed, especially in relation to their behaviour. Similarly the effects of the interaction between new combinations of bee diseases and their vectors need to be the subject of much more extensive research programmes.

In many parts of the world including the British Isles, feral or native bees are less likely to survive than before the arrival of Varroa destructor. This parasitic mite, its association with bee viruses and its deleterious effects, can destroy colonies of feral bees within a few years. This highlights the increasing need for experienced, knowledgeable beekeepers skilled in caring for their colonies as the future of biodiversity and human food security lies partly in their hands. Lack of observational skills leading to poor management, lack of disease and pest control, poor hive maintenance and hygiene and careless manipulations not only lead to the death of individual bees but can result in the loss of the colony.

Many of the reasons why bees die before they reach their full potential can be ameliorated or prevented by competent and experienced beekeepers through Integrated Bee Health Management. The purpose of this book is to describe and explain the elements that are essential to developing IBHM and how to use them, and to encourage those new to the craft to start their beekeeping well informed and with the knowledge of practical and proven methods to help our honey bees.

TWELVE PRINCIPLES OF GOOD BEEKEEPING

1. Take time to observe your bees in the apiary and learn to work with their natural rhythms.

2. Carry out at least four full inspections during your beekeeping season.

3. Keep detailed records.

4. Maintain high levels of hygiene.

5. Practice regular brood frame replacement.

6. Monitor and control Varroa and Nosema.

7. Improve your manipulation skills.

8. Understand how to care for your queens.

9. Re-queen each colony every 2-3 years.

10. Learn to manage swarming.

11. Ensure that your bees have access to good forage throughout the season.

12. Keep up to date with developments in beekeeping and relevant legislation.

INTRODUCTION

Honey bees pollinate flowers which produce seeds and fruits that other animals and birds eat, they are part of the complex food web which is essential to earth's biodiversity.

In the British Isles honey bees are almost totally reliant on beekeepers to ensure their survival. Beekeepers provide the only care and attention they receive, in particular in helping to minimise the impact of the parasitic mite *Varroa destructor* and a number of viruses, diseases and conditions through control measures, including medications that enable them to survive.

For many beekeepers, both new to the craft and those with more experience, today's beekeeping seems to be full of the problems that honey bees have in terms of diseases and conditions. This book aims to show how we can help our bees to resist such problems.

Beekeeping is a year round activity and management practices carried out in one season can affect the colony months later. For example preparations for winter carried out in the autumn will determine the condition of colonies in the spring, whilst spring management will affect the performance of colonies during the summer.

The beekeeper must have knowledge of the natural history of the honey bee and acquire a range of beekeeping skills to keep bees healthy. The following pages describe how these skills and practices can have significant health benefits for colonies. Whether you are keeping bees for honey production, crop pollination, to help maintain biodiversity, or for pleasure, healthy bees are a key factor in achieving good results.

Honey bees, like all animals, are vulnerable to sickness and disease, however as this book will show, good beekeeping practices can improve bee health and minimise the potential of disease to reach a level that exceeds the economic injury level. Disease above this level makes the colony of no economic value and also acts as a reservoir for vectors and the transfer of disease from one colony to another.

The book begins with a short introduction to the natural history of the honey bee as this knowledge is essential in understanding the health status of colonies and in planning and carrying out good beekeeping management practices. Even though beekeepers keep their bees in artificial conditions, honey bees try to fulfil their natural instincts and the most successful beekeepers are those who understand how to work with the natural history and seasonal cycle of the honey bee.

Colony reproduction, more commonly referred to as swarming, is often portrayed as an undesirable phenomenon and yet careful management of this fundamental colony impulse is both good for the colony and good for the beekeeper; this is explained. It is followed by an explanation of the basic needs of the honey bee colony, nutrition and the importance of understanding the total amount of resources the colony will require throughout a year.

The next section introduces the concept and benefits of Integrated Pest Management (IPM) and Integrated Bee Health Management (IBHM). This leads on to the internal and external causes of stress that can be influenced and managed by beekeepers. Stress is a contributing factor to many bee diseases and conditions through a variety of ways including damage to their immune systems and there is much that the beekeeper can do to reduce or even prevent the occurrence of diseases and conditions in a colony by reducing stress.

The diseases of brood and adult bees, including field identification, management and control are described in the next section with particular emphasis on Varroa, followed by a section on pests and other threats to bees.

Healthy bees depend upon good husbandry and sections on how to achieve this follow. The importance of healthy queens is described, together with the management of swarming.

The previous sections introduce the theory of honey bee health and describe methods of dealing with the range of factors that can result in their ill health. They are followed by a section on how this can be translated into practice through an example of a typical year of beekeeping. It illustrates a calendar of operations a beekeeper needs to carry out in response to the colony development through the year.

The final part of the book provides additional detail on a number of necessary and useful techniques to assist beekeepers in managing their colonies, minimising the difficulties that can arise and simplifying some of the more complicated manipulations that are required for good, healthy bees and enjoyable beekeeping.

The reader may find some overlap between sections; this is because all aspects of beekeeping are interrelated. Also we believe that some points are relevant in several different circumstances and others are so important that re-emphasising them is worthwhile.

Training beekeepers

SECTION 1

THE NATURAL HISTORY OF THE HONEY BEE COLONY

In the early 21st century most colonies of honey bees in the British Isles are managed by beekeepers. There are very few wild colonies mainly because of the Varroa mite *(Varroa destructor)* which probably reached the British Isles in the late 1980s and was first identified in England in 1992.

Colonies of the European honey bee *(Apis mellifera)* have uniquely developed the ability to store sufficient food reserves, in the form of honey and pollen, to enable them to survive as a unit of around 10-12,000 bees through the winter when there is little nectar or pollen forage available for them. This over-wintering ability is important for farmers and growers as it provides significant numbers of honey bees for the pollination of early crops in the spring when other pollinating insects are few in number. The contribution made by honey bees in the pollination of crops is of critical importance to human food security and diversity, and the pollination of many other plant species helps to maintain wild plants, and consequently also wildlife diversity. This is often overlooked and yet honey bees, together with other pollinating insects, are essential to human survival.

Honey bees are kept in hives which comprise a number of boxes within which are located the moveable frames on which bees build their combs. The honey bee naturally increases the number of its colonies by swarming, a process whereby the old queen leaves the hive with a proportion of the bees in the swarm to found a new colony, and in the original colony a new queen is produced. Unless caught by a beekeeper the swarm will seek out a new site in which to build combs and establish a feral colony. This may be in a hollow tree, a roof space or similar location where there is sufficient space for the colony to develop. These colonies usually only survive 2-3 years in a feral state before they are killed by the effects of the Varroa mite and the viruses associated with it.

Honey bee colonies are highly organised, matriarchal societies that support a relatively long-lived female (typically 3-5 years) called a queen. She is the mother of all the bees in the colony 95% of which are workers that are always female, they never mate, and never lay eggs as long as the queen is alive and healthy. The other 5% of her offspring will develop into sexual reproductives, either as new queens or males called drones. During the course of a year workers may live between 4 weeks and 9 months depending on the season. A summer colony of around 50,000 bees will lose around 500 bees per day, and over a period of four months this equates to a complete colony change, with the exception of the queen.

Within the colony the queen exercises reproductive dominance over the workers by secreting queen substance that inhibits the development of the workers' ovaries and there is no other line of command. It is believed that each worker bee makes her own decisions determined by her current life phase and this dictates her behaviour, producing small local changes that in turn stimulate other bees to adjust to the new local situation and in turn make their own decisions. The effects of these small changes can be observed in the macro-behaviour of the colony.

Caste development

The queen controls the sex of her offspring through laying either fertilised or unfertilised eggs. Eggs that are not fertilised are haploid, and contain half the diploid number (2n=32) of bee chromosomes, these develop into drones. Eggs that are fertilized contain chromosomes from both the queen and the drones with which she has mated, these are diploid and will develop into either queens or workers. It is the feeding regime given to the larvae that determines their future role in the colony. Larvae which are to become queens are fed royal jelly which has a high hexose sugar concentration. This sugar constitutes about 35% of the royal jelly and is more concentrated than the brood food given to the worker larvae where the sugar concentration is around 10%. The higher sugar concentration, together with a special feeding regime, stimulates a series of physiological and developmental processes to turn them into queens. The feeding regime for queens and workers is complicated and brood food and royal jelly each have different compositions at all stages of larval life.

The three honey bee castes found in the colony.

Queen

Actual size 19mm

Drone

Actual size 17mm

Worker

Actual size 17mm

Photographs are 1.5 times life size

The development of egg / larva / pupa and finally adult phases complete the life cycle of the honey bee in which the organism goes through a complete change of structure in a very short time, this is called metamorphosis. Knowing the time taken in the development and emergence of the adult bees helps in reading the combs when they are inspected by the beekeeper. The presence or absence of eggs, an irregular laying pattern of eggs or larval development, and the presence or absence of drones are all indications of events taking place in the colony. Correct interpretation of the state of the combs, the bees on them and the condition of the larvae will help you to identify abnormalities and decide if you need to take action.

TABLE 1 Development times for each caste

Caste	Days between egg laid and 'hatching'	Days after egg is laid when cell is sealed (prior to pupation)	Days after egg is laid until adult emerges	Days after egg is laid until sexual maturity
Queen	3	8 / 9	16	20+
Drone	3	9	24	34+
Worker	3	9	21	–

Workers and drones are reared in the hexagonal shaped cells of the comb (drone cells are larger than those of workers), whilst the queen develops in an elongated cone shaped cell which normally protrudes and hangs vertically downwards from the face of the comb, with the opening of the cell at its lowest point.

Queen cell hanging from the face of the lower edge of the comb.

Principal Caste functions

The honey bee colony is an amazing society of insects often likened to a superorganism with behavioural interactions and division of labour that are highly complex. A healthy and thriving colony requires the correct balance of each caste appropriate to the time and the natural cycle of the colony throughout the year.

TABLE 2 Principal roles of each caste

Caste	Role
Queen	Reproductive female, lays eggs, provides colony cohesion through her pheromones.
Drone	Reproductive male, produces sperm to fertilise virgin queen on her mating flights, dies very soon after mating.
Worker	Wide range of tasks that are age related, including brood rearing; attending the queen; colony hygiene; wax production and construction of combs; colony defence; foraging and processing of nectar; collection of pollen, water and propolis. Length of life depends on the time of year.

Workers – Division of labour and changes in relation to age

The worker caste carries out the widest range of activities and these are age related. Table 3 shows the typical number of days after emergence for each phase. The timing varies depending upon the state and needs of the colony at the time. If there is a large increase in egg-laying and there are inadequate younger bees to nurse the brood, the older bees may change their behaviour. In the case of a colony suffering from disease or loss of bees, for example due to pesticide poisoning, these timescales may alter and an older bee may revert to her earlier ability to produce brood food in her hypopharyngeal glands to feed young larvae.

TABLE 3 Typical activity phases for worker bees

Time after emergence	Activities
0–6 days	Cell cleaning, general hive cleaning
3–9 days	Producing brood food and feeding the brood
3–15 days	Attending the queen
6–18 days	Nectar and honey processing
12–20 days	Wax production and comb building
15–25 days	Hive ventilation
18–35 days	Guard duty – defending the colony
20 days – death	Nectar collection
20 days – death	Pollen collection
20 days – death	Water and propolis collection

COLONY REPRODUCTION

Pheromones and egg-laying

A pheromone is a chemical, secreted by an exocrine gland. It elicits a behavioural or physiological response by another individual of the same species and acts as a chemical message.

Honey bee colonies normally contain only one queen, and this state is maintained by the interactions of pheromones between the queen, worker brood and workers that influence the behaviour of worker bees and prevent their ovaries from developing, and other responses.

Swarming is influenced by two pheromones produced in the queen's mandibular glands, namely 9-keto-(E)-2 decenoic acid (also known as 9-oxodecenoic acid, 9-ODA, or queen substance) and 9-hydroxy-(E)-2 decenoic acid (also known as 9HDA).

The pheromones produced by the queen contribute towards suppressing the development of the ovaries in worker honey bees, inhibiting their queen rearing activities and the development of queen larvae. The queen also produces a 'footprint' or queen trail pheromone from the tarsal glands located in her feet that is distributed over the surface of the comb throughout the colony wherever she walks. This pheromone suppresses the production of queen cups, an early stage in the rearing of queen cells. In addition worker brood also produces pheromones and these substances exhibit a greater inhibitory effect on worker ovary development than queen pheromones. Workers have the potential to produce functional ovaries, however they cannot lay fertilised eggs, and any eggs laid by workers will develop into drones.

Swarming and the factors that promote it

Swarming is a natural occurrence, it is the main way in which colonies create new queens, spread to different sites and establish new colonies. Understanding swarming and its role in the natural history of the honey bee is an important aspect of honey bee health. For many beekeepers swarming is seen as undesirable, but successfully managing it provides opportunities to create new colonies and can have very beneficial effects on the health of the bees. As swarming is a natural process attempts by the beekeeper to thwart the bees, for example by cutting out queen cells, usually end in failure and the bees are lost in a swarm. It is better to recognise and work with the natural rhythms of the colonies. Before taking out the hive tool and cutting out the first queen cells, decide whether you can use this valuable resource, because queens produced from emergency queen cells may not be of the same quality as the prime queen cells.

The causes of swarming have been studied and observations of colonies made over many years, and there are a number of hypotheses. None of them alone is sufficient to explain swarming and research has shown that swarming initiation is a result of a complex interaction of factors. Two competing hypotheses were developed during the early stages of research on this subject.

The nurse bee or brood food hypothesis was first proposed by Gerstung in 1891 and further developed by Moorland in 1930. Gerstung believed that a surplus of nurse bees in pre-swarming colonies could result in an excess of brood food, and the rearing of queens could be a way of using this up. The second hypothesis, called colony congestion or crowding hypothesis, was proposed by Huber in 1792 and developed by Demuth in 1921. It proposed that the crowding of adult bees, and the resulting limited space for brood rearing, results in the initiation of queen rearing. Neither of the hypotheses was totally satisfactory and it was the discovery of the role of queen pheromones, especially the queen substance 9-ODA, and the transmission of pheromones throughout colonies, that provided a key piece of the jigsaw. Current understanding is that the initiation of queen rearing is due to factors both inside and outside the hive.

The principal internal factors include:
- Brood nest congestion, restricting space for the queen to lay eggs.
- Reduced transmission of queen substance because of crowding and restriction of queen and worker movements.
- A high proportion of young workers.
- The age of the queen. Queens produce less pheromone as they age.
- Genetic traits of the queen.

The principal external factors include:
- The season.
- The weather.
- The geographical area where the bees are kept.
- Conditions of shade, ventilation and temperature.
- The flow of nectar and availability of pollen.

Good queens

There is no doubt that queens produced under the swarming impulse are usually well developed and strong when they are reared from healthy stocks that are good tempered, good nectar gatherers and that produce a small number of queen cells (<10) at swarming time. Colonies produce swarm cells when the nutritional inputs of pollen and nectar are at their best for feeding young workers that subsequently will produce the special food for the queen larvae.

Swarming and Supersedure

The process of swarming by honey bees in the British Isles begins in the middle of winter when colonies start to rear their first worker brood. By April a strong colony will have increased to at least 3 times more than the overwinter population and by May / June this can be 6 times the overwinter numbers, reaching 60,000 or more.

Not all colonies swarm each year and in an average year at least one hive in four may swarm. Colonies with a new queen of less than one year old will have a one in twenty chance of swarming. However like all things in beekeeping nothing is hard and fast. Many non-swarming colonies may replace the original queen with a new queen without a swarm issuing from the hive. This is called supersedure, and it may happen more frequently than we observe. In superseding colonies the old queen and the new queen can coexist and may often be seen on the same comb until the old queen disappears. Colonies with new virgin queens produced by supersedure may cast, but the cast will contain a new virgin queen, rather than the old mated queen.

TABLE 4 Comparison between swarming and supersedure

Characteristic	*Swarming colony*	*Supersedure colony*
Swarm /cast led by	Old queen	Virgin queen
Number of queen cells	Many (1–30)	Few (1–5)
Location on the brood frame	Queen cells hang down from the comb, often from a lower edge, there may be several in a group	Queen cells are isolated, protruding slightly from the face of the comb
Development	Queen cells may be at different stages of development	Queen cells are of a similar age
State of the original queen's health	Healthy	Poor, or the queen is missing
Brood pattern on combs	Compact	Irregular

The usual cause of supersedure may be the diminished production of the pheromone 9-ODA by the queen, but this is not always clear. Colonies may supersede queens that are:

- Injured.
- Diseased, for example infected with *Nosema spp.*
- Laying unfertilised eggs.
- Laying an insufficient number of fertilised eggs.

Queens that are more than 2-3 years old are superseded more frequently than younger ones. Supersedure is most common in the late spring or early autumn, although it can take place any time from early spring until the autumn. About 20% of colonies supersede their queens each year. Supersedure has an adaptive value because it ensures the presence of a laying queen, especially if the virgin queen fails to return from her mating flight. The old queen may continue to lay eggs while the new queens are developing, and often she is not eliminated until the successful virgin queen has mated and begun her own egg-laying.

Colony size may influence whether a colony swarms or supersedes its queen. Large colonies occupying large cavities are more likely to replace their queens by supersedure than by swarming. However colonies that have a strong tendency for supersedure are often found to be queen-less in the spring because they have attempted to supersede their queen late in the season and the new queen has failed to mate. Queens mated in poor weather are frequently superseded.

Preparations for Swarming – reading the combs

The first signs of preparations for swarming can be seen about four weeks before the prime swarm issues from the hive and an early sign is the presence of flying drones and significant amounts of drone brood. Drone rearing generally peaks at this time and a drone becomes sexually mature 37–40 days after the egg is laid.

Queen cups are partly formed queen cells that are found on the comb throughout the period of the year when queen rearing is likely to occur. They are usually found along the tops and bottoms of the brood frames, although cups can be built on the comb face. Some may have an egg laid in them but only a few of these are ever used to rear queens. However, it is always worth checking to see if any of the queen cups have an egg in them in case the workers try to rear a queen from them.

The Swarm

Usually the prime swarm issues from the hive when the first queen cells are sealed (capped). Some colonies will swarm much sooner when only very young larvae are present in the queen cells, others will swarm later when the first virgin queen emerges at around 16 days after the egg was laid. A prime swarm consists of the old queen and 30–70% of the colony's bees. On leaving its hive it usually settles at an interim clustering site.

Casts, or secondary swarms, emerge with virgin queens some days after the prime swarm; each may contain one or several virgin queens. Once a new queen has emerged, the workers that control the whole process may tear down the other queen cells or allow several virgin queens to emerge and coexist and the workers also control the rate of emergence of the young virgin queens. The first virgin queen to emerge may kill the other queens in their cells, usually by stinging them through the walls of the queen cells, unless the workers prevent this from happening.

The original colony now has a virgin queen, a small number of bees and emerging brood. It will be several weeks before new worker brood is produced. The virgin queen may mate quickly and start egg-laying, but bad weather may prevent mating flights and she will become stale and unable to mate. The period for successful mating is about 21 days after the virgin queen has emerged. The colony will die unless the beekeeper intervenes and inserts a comb with worker eggs taken from another colony enabling new queen cells to be raised.

Nest Site Selection

The selection of the new colony site is the last stage of the colony's preparations for swarming; it is usually made 2–3 days before the swarm leaves the hive. Selection prior to swarming is necessary because, although the bees fill up with diluted honey before they issue as the swarm, they must quickly establish a new nest location before they run out of energy. If the weather is cold and wet swarms can soon perish. Scout bees (workers) will search out suitable cavities in unoccupied or bait hives, hollow trees, roofs or chimneys, or any space with an access several metres above ground level in which the colony can establish a new, defendable and weather tight nest site. Bees in a swarm are usually docile, however if the swarm has been out of the hive for a few days the bees can become hungry and bad tempered. When trying to catch a swarm wear appropriate personal protection.

The evaluation of a potential nest site usually takes between a few hours and a week. The scout bees make an examination of the interior and exterior of the potential nest site and hover at increasing distances away from it. The interior examination involves brief flights inside the cavity and rapid movements on the inner surfaces. They may groom each other, exchange nectar and expose their Nasonov glands to each other as part of the site selection process. They will make sporadic returns to the site during the day.

The scout bees periodically return to the swarm cluster and carry out dances on the surface of the cluster that can last 15 to 20 minutes. These dances are similar to the round and waggle dances. The scouts may be evaluating several sites and so the dances will indicate different locations and their desirability. As the evaluation process continues the

A swarm hanging in a tree – relatively easy to catch.

scout bees begin to communicate their preference for the best sites, and the locations indicted by the dances become fewer. Finally the bees appear to reach a consensus when all the scouts dance to indicate the same location.

At this point the scout bees perform a buzzing run over the cluster causing it to break up and take to the air. Guide bees, which may be scout bees, appear to lead the swarm by a combination of rapid flight in the direction of the selected site and by chemical cues, although by the time the swarm takes off many of the worker bees will have read the dances and presumably know the location of the new nest site. The pheromones produced by the queen's mandibular glands play a key role in the swarming process. The pheromone 9–ODA attracts the worker bees whilst in flight, and 9–HDA stimulates the workers to alight on the cluster and stabilise it.

Swarming and its impact on colony health

Brood combs can act as a source of continuing infection for a number of bee diseases such as American Foul Brood (AFB), European Foul Brood (EFB), Chalkbrood and Nosema. When the swarm leaves its parent colony and establishes itself in a new site where there is no wax comb present, it will produce new comb so the chain of infection is broken or significantly reduced. Feral bee colonies resulting from swarming are unlikely to survive more than three years before they succumb to Varroa and its associated viruses, the combs will subsequently be destroyed by wax moth. As long as these unmanaged colonies survive they may be a reservoir of infectivity.

Building new comb

Worker honey bees produce wax through their wax glands and manipulate it to build comb. It is a very energy intense process and it also requires honey bees to have access to good supplies of pollen from which they extract key substances used in the production of wax. A swarm needs to produce about 1200g of wax to build about 100,000 cells and to do this requires some 7.5kg honey.

Wax is a very important resource for the honey bee, but it is also capable of causing health problems if kept in the hive long enough to allow the build up pathogens and undesirable chemical residues such as acaricides. Beekeepers can help to break the chain of disease transmission in their colonies by replacing old brood frames with frames of new foundation, or newly drawn comb, on a regular basis. The changing of brood combs will be discussed in more detail later.

HONEY BEE NUTRITION

Energy requirements

It is only the latter part of the worker bee's life that is spent in foraging activities. The final days of foraging bees may be short in number and so colonies should be placed in situations near to good supplies of forage to enable them to collect the maximum amount of nectar and pollen with the minimum expenditure of energy. The beekeeper's knowledge of the food resource available throughout the active season helps to ensure the development and maintenance of thriving and healthy colonies. Foraging for nectar is about efficiency of collection and the net increase to the colony brought in by the foragers. When expressed on a per kilometre basis the nectar forager brings back some 50 times more energy than was expended in making the foraging flight. A strong colony produces and consumes around 300kg honey in a season and it is estimated that this is produced from the nectar collected in several million foraging trips. Only a small amount of this honey is stored in the hive at any one time because it is being used as fuel so the turnover is high.

Large quantities of energy and protein are required by honey bees to build comb; produce and rear new bees; maintain the colony temperature and forage for nectar, pollen, propolis and water. Not only does the colony need these inputs during the active season, but also requires energy from honey, and protein (from pollen) stored in the combs and in the fat bodies of the over-wintering bees. Honey should not be viewed solely as a food for the maintenance of the bee's body functions, it is the fuel used to produce energy to warm the brood in the summer and keep the clustered bees from freezing in the winter. With this in mind it is important for beekeepers to know the quantity of over-wintering stores needed and to leave this amount in the hive when taking the honey harvest.

As a rule of thumb a colony of bees on a single brood box and a super requires a minimum of 16kg (35lb) and preferably 25kg (55lb) of honey stores and a strong colony on a brood and a half or double brood and super(s) requires at least 27kg (60lb) to successfully overwinter. One BS brood frame holds around 2-3kg (4-6lb) of honey. Honey contains 80% sugars (mainly glucose and fructose) so 4.5kg (10lbs) of honey only contains 3.6kg (8lb) of sugars from which to generate energy. Typically 1kg (2.2lb) of sugar gives around 1.25kg (2.75lb) of honey stores and the balance is made up by the water added by the bees.

Nectar

Honey bees require nectar and pollen to provide carbohydrates, proteins, vitamins, minerals, fats (lipids) and water for healthy, normal growth, development, maintenance and reproduction. Nectar contains 5-80% sugars in water plus small amounts of proteins, minerals, vitamins, lipids, organic acids, pigments, enzymes, volatile oils and pollen grains derived from plants. The main sugars, present in varying proportions, are fructose and glucose (both monosaccharides) and sucrose (a disaccharide). The amount and sugar content of nectar is dependent on the plant species from which it is obtained and environmental factors, especially weather conditions. The nectar is converted into honey by evaporation of water and the addition of enzymes, including invertase (also called sucrase), which convert the sucrose to glucose and fructose. Honey bees will also collect honeydew, which is excreted onto leaves by plant-sucking aphids and shield bugs.

TABLE 5 Typical composition of nectar

Constituent	% by weight
Water	30-90
Sucrose	5-70
Fructose	5-30
Glucose	5-30
Other constituents	up to 2%

More detailed analyses have identified the following substances in samples of nectar, although the composition varies depending on the plant source. Although these substances occur in very small amounts in nectar they are important for good bee nutrition.

- Sugars, such as sucrose, glucose, fructose, xylose, raffinose, melezitose, trehalose, melibiose, maltose, dextrin and rhamnose.
- Vitamins, principally vitamin C and some of the vitamin B complex.
- Amino acids including aspartic acid, glutamic acid, serine, glycine, alanine (the most important to bees of 13 isolated from nectar samples).
- Minerals e.g. potassium, calcium.
- Organic acids e.g. gluconic and citric acids.
- Pigments.
- Aromatic compounds e.g. alcohols and aldehydes.
- Enzymes, including invertase, transglucosidase, transfructosidase, tyrosinase, phosphatases and oxidases.
- Mucus, gums, ethereal oils and dextrin.
- Particulate constituents including pollen, fungi, yeasts and bacteria.
- Antioxidants mainly ascorbic acid (vitamin C).
- Occasionally lipids and alkaloids.
- Very occasionally proteins.

In some years some sources of forage are only available for very short periods or in relatively small amounts. The effect of the weather on the growth and yield of forage, or the weather at the time the bees are trying to forage, can also be limiting. It is important to ensure that the colony has sufficient comb for storage of forage when it is available. Sometimes during the active foraging season the beekeeper may be unaware that the bees are very short of stores because of a reduction in nectar availability and a higher rate of consumption of the stores. Such colonies could be set back to the extent that the workers start to suck the larvae for nourishment, egg-laying stops and consequently, even when foraging resumes, the colony has a big gap in its age profile. It will take several weeks before the foraging force is augmented with new bees. Bees that are poorly nourished are susceptible to disease, so always ensure your colonies have some spare comb capacity in the hives at all times to enable the bees to take advantage of short but good nectar flows.

Processing of nectar and its conversion to honey

The worker bee repeatedly exposes droplets of nectar on her proboscis to reduce the moisture content. She will also 'hang' droplets of the partially processed nectar around the walls of cells. This increases the surface area of the nectar that is exposed to currents of air created by the worker bees fanning and moving the air through the hive. To do this efficiently the bees require comb space. If there is insufficient space, feedback behavioural mechanisms of the bee dance language result in the incoming nectar being processed more slowly; consequently fewer bees are recruited for nectar foraging.

The vital importance of Pollen

Pollen is the major source of protein, minerals, lipids and vitamins for honey bees. A colony requires 20-30kg of pollen per year and it is calculated that its collection requires around 20 million foraging flights.

Protein is essential for brood rearing; for body tissue development; the production of royal jelly/brood food and egg production. The protein status of bees and pollen can be expressed in terms of crude protein, and bees require pollen sources that have a crude protein content of at least 20%. Crude protein values are calculated from the percentage nitrogen content of the protein multiplied by a factor of 6.25. The amount of protein which is bio-available or digestible may be a lower value than that for the crude protein.

Pollen contains the following constituents in varying proportions according to its plant source.
- Lipids-essential for brood food production.
- Carbohydrates & related compounds.
- Major minerals, potassium, sodium, calcium, magnesium.
- Organic acids.
- Free amino acids.
- Nucleic acids.
- Terpenes.
- Enzymes.
- Vitamins, B2 (riboflavin), B3 (niacin), B6 (pyridoxine),

pantothenic acid, biotin; C (ascorbic acid) and E.

- Nucleosides.
- Pigments, carotenoids and flavonoids.
- Plant growth regulators.

TABLE 6 Crude protein content of the pollen of some plant species pollinated by honey bees

Status	Plant Species	% crude protein *
Poor	Sunflower (Helianthus spp.)	13
	Maize (Zea mays)	15
	Weeping willow (Salix spp.)	15
	Lavenders (Lavandula spp.)	20
Average	Pussy or goat willow (Salix caprea)	22
	Oil-seed rape (Brassica napus ssp.oleifera)	24
	Vetches (Vicia spp.)	24
Above average	Almond (Prunus amygdalus)	25
	White clover (Trifolium repens)	26
	Pear (Pyrus spp.)	26
	Gorse (Ulex europaeus)	26
Excellent	Viper's bugloss (Echium vulgare)	35

Data abstracted from Somerville (2003)

*20% is the minimum acceptable crude protein. A correct balance of essential amino acids is also required.

TABLE 7 Mean value of pollen protein content for some plant species of the British Isles

Plant species	Protein content mean % value
Ling *(Calluna vulgaris)*	13.9
Bramble *(Rubus fruticosus)*	15.1
Rosebay Willowherb *(Chamerion angustifolium)*	16.2
Ragwort *(Senecio jacobaea)*	17.2
White clover *(Trifolium repens)*	35.2
Red Bartsia *(Odontites verna)*	36.6
Red clover *(Trifolium pratense)*	40.8

Data from Hanley et al (2008)

The level of body crude protein in honey bees varies from 21% to 67%, and is an indicator of the health of the individual bees and the colony. Ideally colonies should be managed so the average minimum level of body crude protein is above 40%. Workers with a body crude protein level below 40% have a life span of 20–26 days, whilst those with body crude protein content above 40% live for 46–50 days. Worker bees with a body crude protein content of <30% not only have shorter lives, but also are poor honey producers and are more susceptible to diseases such as European Foul Brood (EFB) and Nosema. Bees with high levels of body crude protein, ideally above 60%, will be stronger, live longer and produce more honey than those with lower levels.

Over-wintering bees need high body crude protein levels in the autumn so the colonies can resist Nosema, EFB and Chalkbrood; be able to maintain a strong condition, and enable the colony to build up in the spring. High body crude protein content also provides essential proteins for the development of young larvae at a time when pollen is in short supply and/or forager bees cannot access it. Bees with a low body crude protein in the autumn will not generally overwinter well and will be susceptible to Nosema and spring dwindling.

Honey bee body protein is reduced by honey production; cold or wet weather; wax production and increases in brood numbers, especially during the spring build up period. Bee body protein content is also related to stress. A low stress situation occurs when the hive is producing new bees at a constant rate, the needs of the colony are small, there is little or no nectar to collect, and the ambient air temperature is around

20°C. A high stress level occurs when a heavy nectar flow stimulates both brood rearing and the need for wax for comb space to store and process the nectar and pollen. To cope with this situation bees require surplus pollen with a digestible protein content >20%. A high stress situation may also occur under cool conditions of less than 20°C, or hot conditions when the ambient temperature is above 35°C.

The quality, quantity and variety of pollens are important and bees fed pollen from a range of plants show signs of having a healthier immune system than bees fed with pollen from a single plant source as found in a monoculture. A typical actively breeding colony requires 50-100g pure protein per day. If the pollen sources contain 20% crude protein the bees need to collect 250-500g of pollen per day to satisfy their requirements. Pollen with a crude protein content of 25% requires 200-400g of pollen per day, whilst if the pollen quality is very poor with only 15% crude protein content 340-600g of pollen per day must be collected.

Bees need varying amounts of protein throughout the year. Feeding bees with protein or pollen in late autumn will not stimulate the production of brood unless there is sufficient protein present in the pollen or feed to overcome the bees' seasonal reluctance to produce brood. In spring bees are naturally stimulated by the conditions and will increase their breeding rate even on lower protein content pollen.

Proteins are composed of amino acids some of which are essential for the honey bee. Essential amino acids are required in definite proportions of the protein digested. If one of these amino acids is not at the minimum level required in the protein of a particular pollen a bee cannot utilise all the protein in that pollen. Bees need 10 essential amino acids in quantities ranging from 1 to 4.5% of the protein digested. If one of the essential amino acids is required at 4%, for example iso-leucine, and it is available only at 3%, only 75% of the total protein consumed can be utilised. The practical implication of this is that any bee feed supplements should contain more than 4% of that particular essential amino acid.

TABLE 8 Amino acids essential for honey bee health and their minimum requirements in pollen

Amino acid	Min % required in pollen
Threonine	3.0
Valine★	4.0★
Methionine	1.5
Leucine	4.5
Iso-leucine★	4.0★
Phenylalanine	2.5
Lysine	3.0
Histidine	1.5
Arginine	3.0
Tryptophan	1.0

★ Some pollens may have these amino acids at below the desired levels

Propolis

This is a naturally occurring resinous material that honey bees collect from the leaves and buds of a large variety of trees, shrubs and plants. They use it alone or with beeswax in the construction and adaptation of their nests; to coat the internal surfaces of the woodwork of a conventional beehive like a varnish; to strengthen the comb structure, and as an antiseptic lining in the cells to protect them and the larvae from moulds and other infections. Propolis is stored in significant amounts in the hive.

Water

Honey bee colonies require water throughout the year. During winter and early spring condensation inside the hives is used by the bees to mix 50:50 with stored honey to enable them to use it for food. The average colony requires about 150g of water per day to do this. Later in the year bees regulate the temperature inside the hive by the evaporation of water; this is carried out by older worker bees fanning their wings in unison to create air currents. In hot weather this may mean the bees need to find 1kg water per day.

SECTION 2

APPROACHES TO BEE HEALTH MANAGEMENT

Introduction

The control of pests and diseases and its impact on food security has always been a challenging problem for humankind. All pests are just organisms trying to make a living out of the crops and animals that humans use. Over the last hundred years or so there has been a dramatic increase in the human population and the need for more intensive farming methods to feed it, unfortunately this has resulted in damaging consequences for other forms of life and the environment.

It is well known that our attempts to eradicate or control organisms that damage our interests through the use of chemicals often result in the development of stronger or more aggressive and tolerant strains. As a counter to this we must seek a greater understanding of the biology and life cycles of pests and diseases and their ability to become resistant to chemical control. This knowledge enables us to develop more sophisticated management techniques to use in pest control programmes.

Integrated Pest Management (IPM)

The concept and practice of integrated pest management (IPM) was developed in the 1950s in response to increasing problems with environmental contamination, chemical residues in crops and the emergence of resistant strains in pest populations. IPM programmes became more widely used in the 1960s and 70s and recently their relevance to beekeeping has been recognised, in particular with respect to the control of Varroa and Foul Brood diseases.

IPM programmes are integrated strategies involving techniques based upon the coordinated use of several control methods to minimise chemical inputs, whilst keeping the pest population density below the economic injury level (see definitions on page 34). In other words total elimination of the pest is not necessarily the main aim.

Definitions

- *Pest population density* in the context of bees is a measure of the size of the pest population in relation to the number of bees, for example, the number of Varroa mites per 300 adult bees, or the number of cells infected with Chalkbrood *(Ascosphaera apis)* per 100 cells of brood.

- *Economic injury level (EIL)* is the pest population density threshold above which economic damage is caused.

- *Chemical resistance* is defined as the ability of a pest species (or a strain of the pest species) to tolerate, or avoid death, or a reduction in population when challenged by a chemical control agent.

- *Pest resistance* is the ability of bees to maintain productivity, population growth and colony integrity when infested with a pest.

How IPM can help beekeepers

Viewing pest management from an IPM perspective enables the beekeeper to move from a series of possibly uncoordinated actions to an organised plan to control pests and diseases and ensure healthy bees.

The range of techniques used to control pests in beekeeping can be categorised as follows.

- *Management techniques* such as drone trapping to control Varroa mites.
- *Physical methods* including the selective use of heat, cold, humidity, and light, for example the use of cold to kill wax moth larvae in stored brood comb.
- *Mechanical methods* including destruction such as picking wax moth larvae out of the comb by hand.
- *Barriers*, for example chicken wire against woodpeckers and water traps against ants.
- *Biological methods* that make use of other species including

pathogens of the pests, for example the use of *Bacillus thuringiensis* (Certan®) to control wax moth larvae.

- *Genetic methods* involving the use of breeding techniques that select for pest resistance, and desirable qualities such as the propensity of some strains of honey bees to groom each other and remove diseased bees and larvae from the colony. Such strains are referred to as hygienic bees.

- *Regulation*, whilst forming part of an IPM programme, should always be part of a national bee health strategy so that import restrictions, quarantine and other bee health measures are strictly observed and enforced. This helps in preventing the introduction of pest and disease populations from other countries into the national bee population.

- *The appropriate use of chemicals* including antibiotics and pesticides such as acaricides, and pheromones.

In order to use the techniques of IPM appropriately and effectively it is necessary to evaluate each pest management decision in terms of its impact on the health of your bees, its potential impact on other beekeeper's bees, the chemicals available and the financial viability of your actions.

If you minimise the use of chemicals and ensure that you use them strictly in accordance with the dosing instructions to slow down the rate at which chemical resistance develops you will extend its useful lifespan as a control agent. By reducing the reliance placed on chemicals to control pests, IPM helps to ensure that your hive products are of the highest quality and free from chemical contamination.

Integrated Bee Health Management (IBHM)

IPM is generally restricted in its use to the control of pests and diseases; ensuring a healthy honey bee population requires much more than this.

The term Integrated Bee Health Management (IBHM) is more holistic and encompasses the numerous aspects which interact and will affect the health of bees. Beekeepers are the key to the establishment and maintenance of healthy colonies and they require the knowledge to understand and care for their needs. Careless and ill-informed beekeepers can cause considerable damage and stress to the bees.

Stressors induced by beekeepers

Well nourished bees are usually healthy bees in a strong colony; they are better able to resist the effects of pests and diseases and, from the beekeeper's perspective, more likely to be productive. Some kinds of stress are induced by the beekeeper.

Such stressors include:

- Lack of shelter and poor hive conditions.
- Poor conditions in the apiary such as exposure to wind, cold, and heat.
- Drifting.
- Incompetent beekeeping manipulations.
- Overcrowding of bees in the colony and hives in the apiary.
- Migratory beekeeping.
- Inadequate swarm prevention and control.
- Inadequate food reserves and forage availability.
- Diseases and other health conditions.
- Poor hygiene practices in the apiary; this includes personal hygiene, equipment and clothing.

The interaction of any number of these stressors may make a colony more susceptible to the effects of pathogens and other organisms. Clearly adult bees and larvae showing abnormal signs and behaviour are under some form of stress and unlikely to reach their full potential in terms of colony development. To achieve their full potential the colony must have a well balanced bee age profile, be free from disease and able to fully exploit the forage available. Beekeepers may unnecessarily disrupt the colony and inadvertently cause the death of the queen or the transfer of disease between colonies.

The importance of management and husbandry cannot be over emphasised, inappropriate or careless actions can cause significant stress to the bees. Below are some factors to consider when planning and improving your management techniques, and the health of your bees.

Shelter and inadequate hive conditions

Think about the conditions in which you keep your bees. The careless beekeeper expects them to survive in damp conditions, with entrances at ground level, often blocked by vegetation and with hive entrances as wide as the front of the hive.

In the natural / feral state for honey bees the nest cavity varies in size from 20-80 litres with an average of around 40 litres. In most cases the entrances of feral nests are about 3m above the ground. The entrance area is typically 10-20 cm^2, usually with a southerly facing exposure. Often the bees will modify the nest, for example if it is liable to get wet they will coat the inside of the cavity and plug holes with propolis.

For managed bees in hives there are a number of things you can do to improve their shelter.

- Ensure the hives are weatherproof and fill in any holes with plastic wood filler.
- Place hives off the ground on stands so air can circulate below the floor, this is especially important with solid floors, the circulating air will help to keep the hive floor dry. A better alternative is to provide mesh floors.
- Preferably arrange the entrances to face a generally southerly direction.
- Ensure that entrances are small enough to be defended from attacks by robber bees, wasps and hornets.

Exposure

Exposure to prevailing winds, the increased risk of rain penetrating and soaking the wood of the hive, and consequent chill factors should be considered before deciding on an apiary site and the layout of the hives within it.

Heat can be an even greater threat than cold and all too often one sees hives exposed to the hot summer sun risking killing the bees by heat exhaustion. Bees have to keep the insides of the hives cool otherwise the honey-laden combs will become plastic, deformed and collapse killing the bees by drowning them in honey. Ensure that your hives are positioned where they will be shaded during the hottest part of the day.

Drifting

Care should be taken when placing colonies in the apiary, in particular to prevent bees drifting from their parent colony into adjacent hives. Drifting bees may result in the following taking place:

- Uneven numbers of bees in the hive populations as a result of bees joining another colony.
- The potential for premature swarming preparations to start because of increased numbers of bees in a colony.
- The spread of diseases from one colony to another.
- Colony honey yield measurements being unreliable as indicators of a colony's health and foraging characteristics.

Drifting can be reduced by using painted hives, coloured entrances and local landmarks for bees, such as trees and bushes in the apiary and by placing the hives with their entrances facing in slightly different directions, and as far apart as possible.

Beekeeping manipulations

Beekeeping catalogues offer several varieties of hive tools and if you watch experienced beekeepers you will often find they use different tools for different tasks. The J-tool configuration for example, is good for separating and levering the frames apart and, using the inner curve of the tool, lifting out the frames for inspection smoothly without jarring the bees. The traditional standard hive tool that has a broader, flat but sharper end is more efficient at scraping unwanted comb from the tops and bottoms of the frames. It is also more effective in separating the hive boxes without damaging or splitting the woodwork of the boxes especially at their corners. Such damage will eventually lead to opportunities for robber bees and wasps to gain entrance and steal honey, and possibly prompt the decline and death of the colony.

To improve your manipulation techniques, think about how you are going to handle and turn the frames for inspection. Ensure that you grip the frame firmly and hold it over the open colony so that if you drop it the queen will not be lost. Turn the frames slowly and carefully as rapid movement could throw the bees, including the queen, onto the ground. When taking out tight-fitting combs you can accidently roll the bees up the comb creating a mess of sticky bees and honey. In doing this you not only injure and kill bees, possibly including the queen; potentially spreading Nosema because of smeared abdominal contents, but also risk aggravating the bees and making them more difficult to control.

Sometimes bees will build comb in a way that makes it impossible to lift out the frames without rolling them. Rather than trying to maximise the number of frames in each box, consider using dummy frames which can be inserted in place of one or more frames at the periphery of the boxes. The removal of these dummy frames on starting the inspection of the combs creates a space into which the next frame can be horizontally moved, and then if necessary lifted out for inspection. During the examination the queen may be seen on the frames and then be restrained on the face of the comb using a press-in queen marking cage. This reduces the risk of the queen falling from the comb as you lift it out of the box for closer inspection.

The temptation to investigate what is going on inside a colony can be irresistible. Opening up the hive, taking out and inspecting the frames can disrupt the bees' activities and can be detrimental to them if you do it too frequently. If you want to study your bees by inspecting a colony more frequently dedicate a docile colony specifically for that purpose.

Assessing the mood of your bees before plunging into the colony is very important. Remove the roof and if the bees' faces are looking out of the holes in the clearer board you can expect a reasonable reception. However, if stings are protruding from the tips of their abdomens through the holes, think whether you need to inspect and if you decide to proceed, you will probably need to use your smoker.

Most beekeepers apply smoke at the hive entrance before beginning their inspections, however it is worth thinking about the effect this has on the bees. Honey bees instinctively react to smoke fearing the threat of what could be a forest fire which could destroy their colony. Their reaction is to fill up their honey stomachs ready to vacate the nest site. The smoke excites and may panic the bees, it causes stress and also means a lot of wasted effort in ingesting the honey and then, after the emergency, having to replace it in the comb. However, if you spray water over the entrance of the hive the guard bees react as if it is a passing shower and retreat inside. The colony remains placid and more amenable to being opened up. This is not to suggest that water sprays alone can be used on all occasions; you should always have a lighted smoker ready to help control the bees to prevent them from getting out of hand. When there is a nectar flow on and they are more interested in it than you, a water spray is sufficient.

National hives supported on hive stands to ensure good ventilation, dry floors and easier for the beekeeper to handle.

Many beekeeping books describe techniques that promote the exchange and interchange of combs and brood from one colony to another to equalise colonies, to boost the development of colonies or to improve weak colonies. This practice is very unwise unless you know that your colonies are disease-free. Otherwise you could be speeding up the transmission of disease from one colony to another.

Overcrowding in the colony and the apiary

Colonies kept in inadequate space and unable to expand will invariably swarm. It is important to provide sufficient frames and combs for the expansion of the brood nest and the processing and storage of nectar as honey. Monitor the development of your colonies so you can judge when to add more supers. Adding boxes containing foundation may not help because the bees need to expend time and energy in drawing out the foundation into wax comb, and by the time they have achieved this, the nectar flow may be over.

One of the consequences of overcrowded colonies with large numbers of young bees is the greater risk of the transfer of Acarine mite *(Acarapis woodi)* during its phoretic or free-living phase. This is when it is living outside the bee's body and able to move from bee to bee.

The importance of forage

Evaluate the amount of pollen and nectar sources within foraging reach of the apiary throughout the season, there may be too many colonies in the apiary for the available forage. The loss of forage over the past few decades is extremely worrying for beekeepers. Whilst nectar flower mixes and other improvements for wildlife have been carried out through various EU and national government schemes, for example Environmental Stewardship Schemes in England, the destruction of hedgerows and wild flowers and the decline of crops that are used by honey bees such as Borage and Lucerne has accelerated. The drive to vastly increase food production to feed the world will also impact upon forage for honey bees as well as other insect pollinators including bumble bees. This is not a factor that is within the control of beekeepers, although it may be possible for them to influence farmers, growers and land managers, either directly or through farmers' representatives and the relevant government agency. In the long term unless addressed it may have devastating effects upon beekeeping not only in the British Isles but also in Europe.

Migratory beekeeping

Beekeepers practicing migratory beekeeping frequently move their colonies from one crop to another during the course of a season, sometimes with little regard for the stress that this can cause the bees. If this is not carried out correctly this stress can cause defecation within the hive and if Nosema is present its spread will be accelerated.

Colonies need their hives to be adequately ventilated when they are moved. Without this the bees will be stressed, both by the heat and by their efforts to ventilate the hive. Raised temperatures inside the hive can lead to melted combs and loss of bees. A fine spray of clean water across the travelling screen and onto the tops of the frames will help the bees to regulate the temperature. When moving the colonies aim to travel in the cooler temperatures of early morning or late evening.

Often colonies are moved from one crop to another with supers containing newly extracted combs. Because the bees will not have had time to secure the frames to the hive body with propolis, their spacing can become displaced during the journey, and they may be jolted out of position. Good practice is to load the hives onto the trailer with the frames lined up in the direction of travel to help prevent this.

Inadequate swarm prevention and control

Many beekeepers do not understand the reasons why bees swarm and what triggers it and this can diminish their enjoyment of beekeeping. By studying bee behaviour and considering how to work with the rhythm of the colony, the beekeeper and thus the bees will benefit. Section 7- The Management of Swarming will help the beekeeper to become more confident and skilful.

Inadequate food reserves and beekeepers' greed

We have seen how much food a colony requires during the active season and over the winter when the weather is unsuitable for foraging. Before stripping the colonies of all their honey stores think carefully about the wisdom of doing this. Honey is the best food for winter stores, provided it is not a type that becomes granulated into very large crystals which the bees cannot process. Feeding syrup to make up for the essential stores taken by the beekeeper is expensive, it takes time and often cold weather will prevent the bees from taking down the syrup and processing it for storage.

Diseases and conditions

Beekeepers need knowledge and information to enable them to manage bees to reduce the risks of diseases and conditions, this includes being able to recognise the signs that something is wrong in the colony and must be investigated. Good quality information and support is available from the relevant government agencies and national and local beekeeping associations. Wherever possible attend courses where you can actually examine diseased combs of brood and adult bees under laboratory conditions with a specialist in bee diseases to advise you. The signs, symptoms, monitoring techniques, methods of treatment and control of bee diseases typically encountered in the British Isles are covered in Section 3.

Some diseases are the subject of statutory control and their status can change over time. Beekeepers need to be aware of the latest position on the status of a disease in their country and their legal obligations about notification. Publications including excellent pictures and descriptions are available from the relevant government agency and beekeepers should ensure that they have the latest editions of them.

Hygiene

Beekeepers can increase the rate of spread of a disease through using contaminated equipment such as dirty hive tools, gloves and bee suits. Similarly leaving old combs or discarded wax in the apiary for the bees to clean up can also spread disease. Good practices that help to reduce the potential for transfer of disease because of poor hygiene are described in the section on keeping equipment clean Section 9 – Honey House Cleanliness.

Strains of honey bees vary in their degree of hygienic behaviour in terms of hive cleanliness, continual cleaning of the queen, cleaning of nursery cells before the eggs are laid, the rapid removal of diseased and dead bees and larvae as well as mutual grooming activities. The qualities of queens in such colonies are well worth selecting for rearing new queens from them.

SECTION 3

HONEY BEE DISEASES AND CONDITIONS

This section covers those aspects of the diseases and conditions that beekeepers can influence through integrated bee health management. In particular it aims to enable beekeepers to recognise the signs and symptoms of diseases and conditions, and the actions they can take to manage them, and to practice beekeeping to reduce the potential for their bees to become diseased, or subject to debilitating conditions.

The legislation involving the status of various bee diseases changes with time and the speed and distribution of the disease through the British Isles bee population. When new pests, diseases or vectors of disease are first indentified in the British Isles the beekeeper may be required by law to notify the authorities if he or she has identified the signs and suspects that the pest or disease may be present in their colonies. This is because the new pest or disease is likely to be statutorily notifiable, for example Small Hive Beetle *(Aethina tumida)*. As it becomes endemic or widespread this may change and beekeepers should check on the legal status of the pest, disease or condition and their legal obligations. This can be done through their local, county or national beekeeping associations as well as through the government agency responsible for honey bee health and inspection. Beekeepers are advised to periodically check the information provided by their bee health authorities, determine their obligations and what assistance they can expect to receive.

BROOD DISEASES AND CONDITIONS, DIAGNOSIS, TREATMENT AND CONTROL

Beekeepers should be aware of all the possible diseases and conditions that may affect their bees. Not all of those referred to in this book will be present in your country at the time of reading, and not all countries have the same treatment and control methods.

The following tables describe the signs shown by the main brood diseases that may be found in a hive. However, sometimes confirmation of the diagnosis has to be made in a laboratory. The tables also describe the management techniques that can be used to help prevent and control these diseases and conditions. (continued on page 59)

TABLE 9 Signs, symptoms, treatment and management of honey bee brood diseases compared with healthy brood

Brood Disease or Condition	Sign/Symptom	Treatment/Management
Healthy brood	• Uncapped cells, pearly white, 'C' shaped larvae, body segments clearly visible. • Capped cells, uniform light brown colour, dry, slightly convex or domed cappings sealing the cells.	• Understand how your bee management is achieving this positive position and maintain it.
American Foul brood (AFB) (*Paenibacillus larvae subspp. larvae*)	• Usually affects brood only at the sealed brood stage but some strains of AFB will kill uncapped larvae. • Moist, dark, sunken perforated cappings. 'Pepper pot' or mosaic brood pattern. Scales (dried larval remains), which are difficult to remove, can be found on bottom walls of open cells. • Most brown decomposing larvae will rope when using the matchstick test so the test is not always diagnostic.	• In the UK bees, frames and combs are destroyed by fire under official supervision and the hive bodies sterilised by scorching with a blowtorch. • Management techniques include regular brood comb changes and use of the artificial or shook swarm technique. Ensure you use the correct technique to check for the presence of scales in the bottom of the cells.
European Foul brood (EFB) (*Melissococcus plutonius*)	• Affects mainly unsealed brood. Infected larvae are discoloured yellow-brown, lying in abnormal positions in the cell. They have a 'melted' appearance. Some dark sunken cappings may be present, but the cell contents will not form a rope with the matchstick test. • Dead larvae. • Form scales that are easily removed by bees.	• Lightly infected colonies may be treated with antibiotics under official supervision and / or use of the shook swarm technique. • Severe cases are usually destroyed by fire as with AFB. • Management techniques include regular brood comb changes and artificial or shook swarm technique.
Chalkbrood (*Ascosphaera apis*)	• Only affects sealed brood. Cells have perforated cappings containing hard, white or mottled grey chalk-like remains that resemble mummies. • Infected colonies rear fewer drones. • Mummies can easily be dislodged from the combs by tapping the frame against a hard surface.	• No specific treatment. • Avoid manipulations which will result in colonies having too much brood to rear early in the spring. For example over-stimulating brood rearing. Avoid using artificial swarm control techniques too early in the season.

Condition	Signs	Treatment / Management
(continued)		• Keep strong colonies. Re-queen severely affected colonies. • Ensure hive is well ventilated, e.g. use a mesh screen floor. • Do not move frames from infected colonies to non infected ones.
Sacbrood (*Sacbrood virus*)	• Only affects sealed brood and larvae die in pre-pupa stage. • Perforated cappings. • Larvae become yellow, then dark brown in fluid–filled sacs 'Chinese slipper' and die after the cell is capped, i.e. larva fails to pupate. • Watery contents of the cells will not rope.	• No specific treatment. • Re-queen severely affected colonies. • Often rectifies itself once a good nectar flow begins. • Manage by regular brood comb changes.
Black Queen Cell virus	• Associated with Nosema. • Affects the queen bee pupae forming a black ring around the tip of the queen cell. • The queen usually dies and turns black. • Any queens that do hatch are of poor quality.	• Manage and control Nosema. • Re-queen the colony. • Management techniques include regular brood comb changes and artificial or shook swarm technique.
Slow Paralysis virus	• Dead shrivelled larvae in sealed cells.	• Manage and control Varroa.
Queen Half Moon Syndrome	• Larva dies before capping, in a twisted letter C or half moon shape in the cell. It changes from white to yellow to light brown and then dark brown with their tracheal lines still present. • Many cells contain multiple eggs with the eggs in chains on the wall of the cell. • Drone eggs are often laid in worker cells. • The queen may be superseded.	• Caused by poor nutrition after emergence.
Bald brood When specifically caused by larvae of Greater Wax Moth chewing through the cappings revealing the head of the larvae.	• Abnormal cappings over worker cells. • Affected cells have a round hole in the capping sometimes with a slight protrusion. • Damage is usually in straight lines of cells. • Pupae have normal appearance. • Signs of wax moth larvae may be visible in the comb. • Larvae usually pupate and emerge normally.	• No specific treatment. • Control wax moth infestation.

TABLE 9 Signs, symptoms, treatment and management of honey bee brood diseases compared with healthy brood *continued*

Brood Disease or Condition	Sign/Symptom	Treatment/Management
Drone laying queen	• Domed drone cappings over worker cells. • Abnormally small drone pupae within cells. • Single eggs are laid in cell bottom. • Unsealed brood may be neglected and dying.	• Replace drone laying queen.
Laying workers	• Domed drone cappings over worker cells. • Abnormally small drone pupae within cells. • Multiple eggs may be present adhering to sides of the cell. • Unsealed brood may be neglected and dying.	• Move colony 20m from its position in the apiary. • Shake bees off each comb onto the ground and allow them to find their way into other colonies in the apiary. Only do this if you know health status of the colony, otherwise kill the colony.
Chilled brood	• Dead brood usually present in all stages, and usually at the edge of the brood nest. • Unsealed brood turns very dark brown or black in colour before drying up.	• Avoid conditions in which bees are unable to care effectively for the entire brood. • Nuclei made up and kept in the same apiary often lose too many bees back to their original colony. See Section 9 – producing and using nuclei. • Spreading of the brood or isolation of brood may also result in chilled brood.
Varroa (*V. destructor*) infestation (Parasitic mite syndrome)	• Signs vary considerably. Sealed brood may be partially uncapped, or with sunken perforated cappings and twisted C– shaped larvae. • Dead larvae / pupae are discoloured brown or black, watery or firm, but never ropey eventually drying to a scale similar to EFB which can easily be removed. • Queen is often superseded.	• Control Varroa infestation to achieve low population levels using appropriate control methods and treatment.

	Symptoms	Control
Small Hive beetle (*Athena tumida*)	• Beetles and larvae may be found in comb cells and the corners of frames. • Maggot-like larvae distinguishable from wax moth larvae by having spines on the dorsum (back) of the larvae and three prolegs near the head of the larva. There may be up to 30 larvae per cell and there is no webbing or frass (debris) as with wax moth infestation. • Combs have slimy appearance and smell of rotten oranges.	• If Small Hive Beetle reaches the British Isles it is likely that the control methods used would include treatment of colony and treatment of soil in the apiary with an authorised contact insecticide under official supervision.
Asian bee mites (*Tropilaelaps clareae and T. koenigerum*)	• Damage to brood similar to Varroa resulting in irregular brood patterns.	• Control infestation to low levels using appropriate control methods and treatment.
Genetical faults **Bald Brood**	• Large areas of brood with no cappings.	• Introduce new queens.

Plate 1 - Developing queen cells, sealed and unsealed worker brood, sealed drone brood, pollen, honey / nectar can all be seen in this picture.

Plate 2 - Healthy sealed worker brood.
The empty cells may be used by heater bees.

Plate 3 - Healthy comb of unsealed worker larvae.

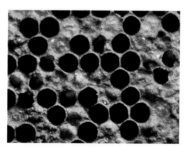

Plate 4 - The sunken cappings on comb infected
with American Foul Brood (AFB). The larvae in these cells will be dead.

Plate 5 – The scales inside these cells are hard to remove, typical of
American Foul Brood (AFB). They are the remains of dead larvae.
To see them hold the comb facing the light and turn the frame in order to
illuminate them in the bottom of the cells.

Plate 6 – Using a match to carry out the rope test for AFB.

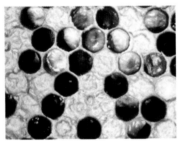

Plate 7 – Dying larvae in uncapped cells;
one of the signs of European Foul Brood (EFB).

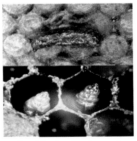

Plate 8a – Larvae infected with the Sacbrood virus (Chinese slipper).
Plate 8b – Dead larvae after uncapping by worker bees,
subsequently the bees will remove them.

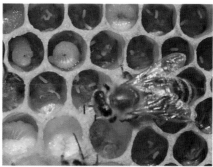

Plate 9 – Multiple eggs in cells can be a sign of laying workers (eggs adhering to the sides of the cells) or a very young queen (multiple eggs laid on the base of cells).

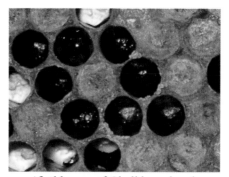

Plate 10 – The mummified larvae of Chalkbrood and empty cells where the mummies have been removed by the bees.

Plate 11 – Chalkbrood mummies on the floor of the hive.

Plate 12 - The domed cappings of drone brood.
Drone brood raised in modified worker cells indicates a drone laying queen.

Plate 13 - Larva of Greater wax moth on comb.

Plate 14 - Adult wax moth on comb.

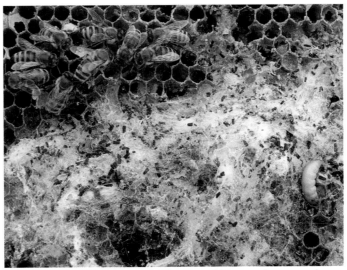

Plate 15 – Silken tunnels, faeces and larvae of wax moth

Plate 16 – Adult female varroa mite on drone honey bee larva.

Plate 17 – Varroa mites of differing ages on drone pupa.

Plate 18 – Adult Death's-Head Hawk-moth

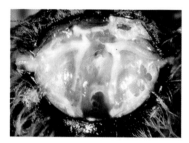

Plate 19 – Dissection for the Acarine mite
showing healthy uninfected tracheae.

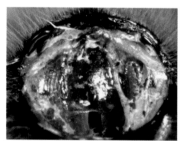

Plate 20 – Dissection for the Acarine mite showing stained
tracheae indicating an infection.

Selected Reading

Aston, D. and Bucknall, S.A. (2004) *Plants and Honey Bees – their relationships.* (Northern Bee Books)

Bailey, L. and Ball, B.V. (1991) *Honey Bee Pathology* Second Edition. (Academic Press)

Caron, D.M. (1999) *Honey Bee Biology and Beekeeping.* (WICWAS Press)

Coggshall, W.L. and **Morse, R.A.** (1984) *Beeswax. Production, Harvesting, Processing and Products.* (WICWAS Press)

Cooper, B.A. (1986) *The Honeybees of the British Isles.* (British Isles Bee Breeder's Association BIBBA)

Crane, E. (1990) *Bees and Beekeeping. Science, Practice and World Resources.* (Heinemann Newnes)

Dews, J.E. and Milner, E. (1993) *Breeding Better Bees using simple modern methods.* (British Isles Bee Breeder's Association BIBBA)

Hanley, M.E. et al. (2008) *Breeding System, pollinator choice and variation in pollen quality in British herbaceous plants.* (Functional Ecology 2008, 22, 592- 598)

Johansson, M.P. and T.S.K (1978) *Some important operations in bee management.* (International Bee Research Association IBRA)

Laidlaw, H.H. and Page, R.E (1997) *Queen Rearing and Bee Breeding.* (WICWAS Press)

Lodesani, M. and Costa, C. editors (2005) *Beekeeping and conserving biodiversity of honeybees - sustainable bee breeding – theoretical and practical guide.* (Northern Bee Books for BABE Beekeeping and Apis Biodiversity in Europe)

Otis, G.W., Wheeler, D.E., Buck, N. and Mattila, H.R. (2004) *Storage Proteins in winter honey bees.* (APIACTA 38), pp 352-357

Rhodes, J. and Sommerville, D. (2003) *Introduction and early performance of queen bees – some factors affecting success.* Publication Number 03/049 (Rural Industries Research and Development Corporation), NSW Department of Agriculture, Australia

Seeley, T.D. (1985) *Honeybee Ecology.* Princeton University Press, (Princeton)

Seeley, T.D. (1995) *The wisdom of the hive. The social physiology of honey bee colonies.* (Harvard University)

Sims, D. (1997) *Sixty Years with Bees.* (Northern Bee Books)

Sommerville, D.C. (2001) *Nutritional value of bee collected pollens.* Publication Number 01/047, (Rural Industries Research and Development Corporation), NSW Department of Agriculture, Australia

Sommerville, D. (2005) *Fat Bees – Skinny Bees, a manual on honey bee nutrition for beekeepers.* Publication Number 05/054, (Rural Industries Research and Development Corporation), NSW Department of Primary Industries, Australia

Tautz, J. (2008) *The Buzz about Bees. Biology of a Superorganism.* (Springer-Verlag)

Winston, M. (1987) *The biology of the honey bee.* (Harvard University Press), Cambridge Mass.

Woodward, D. (2007) *Queen Bee: Biology, Rearing and Breeding.* (Northern Bee Books)

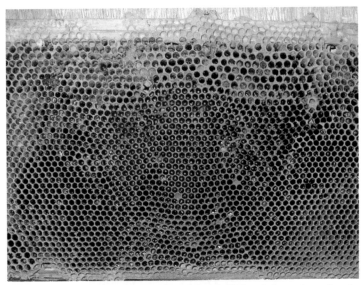

Plate 21 – When a brood frame reaches this condition it is ready to be changed. Then dispose of it, preferably by burning.

Plate 22 – Old mouldy comb should also be removed from colonies and destroyed.

Plate 23 – Adult Small Hive Beetle surrounded by worker bees.

Plate 24 – Lateral view of adult Small Hive Beetle.

Plate 25 – Small Hive Beetle larvae (see centre of picture)
will uncap and eat bee brood and honey and stored pollen.

TABLE 10 The spread of brood diseases

Sources of infection	Means of spread
Infected combs, brood combs	Transfer of combs between colonies.
Super combs	Robbing. Transfer of combs between hives.
Honey containing pathogens	Drifting and robbing.
Beekeeping equipment	Poor management practices. Poor hygiene with bee suits, veils, hive tools and hive parts.

Field diagnoses of larvae suspected of being infected with AFB or EFB can be made using special diagnostic kits called lateral flow devices which utilise antibody technology specific to each bacterium species; these can be obtained from beekeeping equipment suppliers. The principle of the device is to disclose AFB or EFB specific antibodies on a nitrocellulose membrane; the presence of a blue line across the test kit window confirms the presence of the bacterium. The kits can be used by beekeepers however some training on their use and the interpretation of the results is advisable. This training can be obtained from your local bee inspector.

ADULT BEE DISEASES AND CONDITIONS, DIAGNOSIS, TREATMENT AND CONTROL

The following table describes the signs shown by the main diseases of adult bees that may be found in a hive. The table also describes the management techniques that can be used to help prevent and control these diseases and conditions.

TABLE 11 Signs, symptoms and treatment and management of adult honey bee diseases

Disease or condition	Sign/Symptom	Treatment/Management
Acarine (*Acarapis woodi*)	• Few colonies actually become severely infested. • Lives of infested overwintering bees are shortened. • It is associated with acute and chronic paralysis virus.	• There is no specific treatment. • Keep strong colonies and minimise risk of robbing.
Nosema (*Nosema apis* and *Nosema ceranae*)	• Some strains of *N. ceranae* are known to kill bees. • No outward symptoms but shortens the life span. Diagnosis involves microscopic examination. • Typically reduces colony population in late winter and early spring. • *N. apis* accelerates the rate of behavioural development in honey bee workers but the underlying mechanisms are unclear. • Underdevelopment of hypopharyngeal glands in infected bees. • Reduction in the production of royal jelly. • Infected queens lay fewer eggs and are more likely to be superseded and lost 10–15 days after ingestion of the Nosema spores. • Last eggs laid by infected queen often shrivel and fail to hatch making it difficult for the worker bees to rear a replacement queen. • Associated with Black Queen Cell virus and Kashmir bee virus.	• Nosema infection, both *N. apis* and *N. ceranae* can be reduced by feeding with the antibiotic fumigillin (Fumidil®B) given in sugar syrup. • Combs can be sterilised using fumes of 80% acetic acid. • Management techniques include regular brood comb changes and artificial / shook swarm technique to reduce increase of resistant and long living spores in the combs which will result in re-infection. • There is evidence that thymol also has some effectiveness against both *Nosema spp*. Products such as Api Herb and Vita Feed Gold are also thought to be effective in the control of *N. spp*.
Amoeba (*Malpighamoeba mellificae*)	• There are no specific symptoms and infected bees appear normal. • Usually only identified by dissection, e.g. when carrying out a Nosema smear.	• No specific treatment. • Combs can be sterilised using fumes of 80% acetic acid.

Condition	Symptoms	Treatment / Management
Dysentery	• Faecal spotting and soiling by adult bee faeces over the combs and the inside and outside of the hive. • Caused by excessive water accumulation and / or feeding contaminated food or unsuitable winter stores. • Dysentery is not a disease, but a symptom of a disease or nutritional disorder.	• No specific treatment available. • Destroy soiled combs. • Dysentery is exacerbated by extended confinement in the hives in bad weather, fermented honey stores, and acid-inverted sucrose. • Complete feeding in time to allow bees to process the syrup into stores before cold weather sets in. Only feed with refined sucrose or table sugar. • Management techniques include regular brood comb changes and artificial / shook swarm technique.
Varroa *(Varroa destructor)* Parasitic mite syndrome	• Stunted adults with deformed wings and shrunken abdomens. • Symptoms include a reduced hive population, crawling bees at the hive entrance and queen supersedure. • Drones emerging from Varroa infected pupae often appear normal but the majority cannot fly and only around 20% reach sexual maturity. • The amount of sperm produced is halved if the drone emerges from a pupa infested with more than 3 mites. • Associated with slow paralysis (SPV) and deformed wing virus (DWV).	• Control mite infestation to low levels using appropriate control methods and treatment.
Chronic paralysis bee virus (CPBV)	• Causes abnormal trembling in adult bees, some paralysis that limits flight (crawling bees on the ground and up grass stems) and bloated abdomens. • Infected bees at all times have reduced hair cover and appear black and shiny.	• Destroy the colony and burn all bees and combs. • Do not reuse combs from infected colonies. • Sterilise the inside of the hive by scorching with a blow torch.
Acute paralysis bee virus (APBV)	• Varroa is a vector. • Manifestations are similar to CPBV, but once infected bee death is rapid.	• Destroy the colony and burn all bees and combs. • Do not reuse combs from infected colonies. • Sterilise the inside of the hive by scorching with a blow torch. • Keep Varroa population under control.

TABLE 11 Signs, symptoms and treatment and management of adult honey bee diseases *continued*

Slow paralysis virus (SPV)	• Infected bees can die very quickly once infected.	• Keep Varroa population under control. • Treat as below
Deformed wing virus (DWV)	• Varroa is a vector. • Infected bees have deformed or poorly developed wings and die from the virus.	• Keep Varroa population under control. • Destroy the colony and burn all bees and combs. • Do not reuse combs from infected colonies. • Sterilise the inside of the hive by scorching with a blow torch.
Cloudy wing virus (CWV)	• Varroa is a vector.	• Destroy the colony and burn all bees and combs. • Do not reuse combs from infected colonies. • Sterilise the inside of the hive by scorching with a blow torch.
Kashmir Bee virus (KBV)	• Varroa is a vector.	• Destroy the colony and burn all bees and combs. • Do not reuse combs from infected colonies. • Sterilise the inside of the hive by scorching with a blow torch.
Asian bee mites (*Tropilaelaps clareae and T. koenigerum*)	• Stunted adults with deformed wings and shrunken abdomens.	• Control infestation to low levels using appropriate control methods and treatment.
Genetical faults White–eyed drones	• Presence of drones with white eyes.	• Introduce new queens.

NOSEMA SPECIES

Good beekeeping practices can help to restore the health of a colony suffering from Nosema, a parasitical spore-forming organism (Microsporan) causing a disease called nosemosis. There are two species of Nosema found in the honey bee in the British Isles, *Nosema apis* and *N. ceranae*. Identification of both Nosema species can be made by microscopical examination of an aqueous suspension of macerated bee abdomens or by sophisticated biochemical techniques.

The parasite invades the digestive cells lining the mid gut of the bee, feeding on the contents of the gut cells and rapidly multiplying until, after a few days, the cells are full of spores. These spores are the resistant resting stage of Nosema. When the gut cells rupture and cause lesions in the epithelial lining of the ventriculus the spores are shed into the gut where they accumulate and are excreted. In winter and spring the spores in the faeces are shed onto the comb faces, whilst in the summer most of them are voided when the bees are on the wing. If the spores from the faeces are ingested by other bees, especially young bees when cleaning cells and comb faces, they can germinate and initiate another cycle of infection and multiplication. Bees infected with *N. ceranae* may appear symptomless, but observation has revealed that they die more quickly compared to infection with *N. apis*. Adult worker bees infected with Nosema die away from the colony.

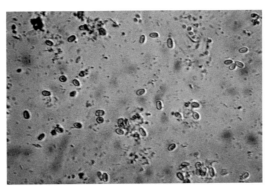

Microscopical slide of a smear of honey bee gut
contents showing spores of Nosema

Queens infected with *N. ceranae* have been shown to die about three weeks after they become infected by diseased worker bees in their colony. Colonies may recognise that the queen is ailing and attempt to replace her by supersedure and this may partly explain why new queens introduced into colonies are often replaced by supersedure, if the attempt at supersedure fails the colony becomes queen-less.

Nosema may facilitate the access of bee viruses through the perforations it makes in the peritrophic membranes and gut epithelium. These perforations allow access for the viruses, carried by Nosema, to the haemolymph of the bee, thereby bypassing the bee's natural defences. Beekeepers need to recognise the importance of controlling Nosema.

Factors affecting the spread of Nosema

- The spores are highly resistant to temperature and chemical agents, and can remain viable in the comb for several decades.
- Infected bees may be squashed or smeared during frame manipulation thus spreading the spores.
- Infrequently changed brood combs lead to a build up of spores in the combs increasing the chance of infection.
- Brood combs from infected colonies can inadvertently introduce the disease into healthy colonies.
- Both *N. apis* and *N. ceranae* can be transferred on hive equipment.

Methods of control and treatment

Monitoring of your colonies enables you to assess whether the colonies are slower to build up than expected. Look for signs of comb fouling or fouling above the hive entrance. If you do not have a microscope and the skill to confirm Nosema, find a competent person to do it; ask your local beekeeping association, there may be a small charge for this service.

Currently both species of Nosema are controlled by the same antibiotic substance called Fumagillin, of which Fumidil®B is a proprietary product available from beekeeping equipment suppliers. Read and follow the instructions for use. This control of Nosema also reduces the vectoring of a number of viruses.

Once Nosema is confirmed carry out the following procedure:
1. Ensure that your bee suit, gloves and hive tools are clean.
2. Use dummy frames to give you more space to manipulate the combs.

3.   Take care to prevent rolling and squashing the bees which will release their potentially infective gut contents.

4.   Transfer the colony onto fresh foundation and feed using syrup containing Fumidil®B. (Follow the instructions on the packaging).

5.   Do not reuse old combs from colonies that have died out and do not exchange combs between colonies.

ACARINE *(Acarapis woodi)*

Acarine is a parasitic mite that infests the breathing tubes, or tracheae of the adult honey bee and causes the disease called acarapisosis (acariosis). It can affect queens, workers and drones. Its incidence can increase when there are sudden very large numbers of young bees in a colony and the risk of transfer of the mite from infected adult bees to the more vulnerable newly emerged bees is greater. Currently there is no authorised treatment for Acarine mite in the British Isles and control measures rely on monitoring the colony, preventing robbing by other bees, and maintaining strong healthy colonies. However, it is thought that products containing thymol may help in controlling Acarine.

VARROA *(Varroa destructor)*

Varroa is a parasitic mite that infests both larvae and adult bees and causes the disease called varroasis. Scientific research has enabled strategies for the control of Varroa to evolve rapidly. In recent years the most significant development has been the recognition that it is important to control Varroa populations in colonies for several bee health related reasons. For example the control of Varroa is crucial in helping to reduce the activation and impact of a number of viruses on the health of the colony. Varroa has the effect of suppressing the bee immune response, as well as providing a means whereby viruses can bypass the honey bee's natural defences, for example the tough exoskeleton, and pass directly into the bee's haemolymph.

Colony collapse

If the Varroa population is allowed to increase without any control measures being applied the colony's well being and social cohesion begins to break down and eventually the colony collapses. In the British

Isles this seems to take place most commonly in August and September but it can also occur in March, April and May.

The signs of colony collapse include:

- A sudden decrease in the adult population, usually with only a few dead bees found in the hive.
- Varroa mites readily seen on the surviving bees, or on any remaining pupae and the hive floor.
- Bees with deformed wings and abdomens.

BEE VIRUSES

Viruses are found in all life stages of the honey bee and have also been found in semen and eggs. Research has shown that viruses can be transmitted horizontally via contaminated food such as brood food, honey, pollen and royal jelly. They can also be transmitted vertically e.g. from queens to their eggs, and the presence of virus in semen suggests that some viruses may be transmitted during mating.

Under normal conditions most honey bee viruses exist in a temporarily inactive state and do not cause signs of disease. Viruses require a vector to transmit them and the Varroa mite has been identified as an important vector in the spread of viral diseases in honey bee colonies. Viral replication is activated by the presence of Varroa and it is also thought to have changed the way in which the viruses enter the bee's body, thereby circumventing the bees' natural defences. Clearly the control of Varroa in honey bee colonies is of utmost importance.

Some diseases and their associated viruses can be transmitted through contaminated honey and pollen so it is good practice to minimise their spread by not using combs from colonies that have died out to feed other colonies. These combs should be destroyed by burning and the hive bodies sterilised inside by scorching with a blow torch. It is very difficult to treat honey bee viruses; an effective approach is to treat their vectors and to identify and understand the risk factors and find ways to reduce exposure to them.

Currently identified bee viruses include:

Acute Bee Paralysis Virus (ABPV)

Bee Virus X

Bee Virus Y

Black Queen Cell Virus (BQCV)

Chronic Paralysis Virus (CPV)

Cloudy Wing Virus (CWV)

Deformed Wing Virus (DWV)

Filamentous Virus

Israeli Acute Paralysis Virus (IAPV)

Kashmir Bee Virus (KBV)

Sacbrood Virus (SBV)

Slow Paralysis Virus (SPV)

Adult worker bee with deformed wings.

THE MANAGEMENT AND CONTROL OF VARROA

Initially good control of Varroa was achieved by inserting into the hive impregnated strips containing pyrethroid-based acaricides such as Bayvoral® and Apistan®. However, the efficacy of these chemicals is being compromised by the development of pyrethroid-resistant Varroa mites. This is not to say that these chemicals are not useful, but in order to reduce the potential for Varroa to develop widespread resistance, the authorised acaricides should be used as part of an integrated bee health management strategy i.e. including monitoring mite levels at regular intervals, using biotechnical control techniques and in some cases using other chemical substances. The legal status of these substances e.g. oxalic acid, lactic acid and formic acid is, at the time of writing, under discussion and may change. Beekeepers should check regularly with the bee health authorities and those responsible for veterinary medicines for their agreement that such substances can be used for the control of Varroa, and determine whether there are any particular restrictions in their use.

Monitoring your colonies

Do not be complacent about Varroa just because your colonies appear to be strong and are producing a good honey crop.

To ensure that your colonies continue to be healthy and productive, it is vital to be aware of the number of mites in each colony and to manage them accordingly. For many beekeepers an awareness of the presence of the mite, followed by treatment with Bayvoral® or Apistan® in the autumn has been sufficient to control Varroa, but the spread of pyrethroid-resistant mites will probably continue throughout the British Isles requiring a change in how Varroa is managed and controlled.

Monitoring times and programmes

All beekeepers should implement a mite-monitoring programme, and monitoring their colonies at least four times a year is recommended. Firstly in the late winter / early spring when increasing amounts of new brood in the colony will provide ideal breeding conditions for the mite; then after the spring flow when the growing colony contains a large amount of brood, especially drone brood, and again in the summer (July and August) to see if mite numbers have built up to dangerous levels. Finally monitor in the autumn when colonies are vulnerable to damage affecting the over- wintering bees that are vital for the survival of the colony. Good colonies can suddenly collapse as a result of an invasion of mites carried on bees from disintegrating colonies. In parts of the country where there is no break in the brood cycle during the winter, monitoring in the winter months is also advisable.

MONITORING TECHNIQUES FOR VARROA

Most beekeepers are always short of time, so techniques that take up little time are valuable. Different techniques have differing degrees of reliability and sensitivity. However, just using one coupled with good observation when examining your colonies, e.g. noting bees with deformed wings, is much better than doing nothing. Visual inspection of adult bees cannot be relied upon solely because the Varroa mites often hide between the plates of the abdomen, and because they run away from the light very quickly you are unlikely to see bees with mites on them unless the Varroa population is high.

If you have a large number of colonies >10 it is impractical to sample them all. Apiaries should be considered as entities so it may be appropriate to select a few colonies and monitor them closely, applying the results to all the colonies in your apiary. It is recommended that very large and very small colonies, and colonies at the ends of rows that may pick up drifting bees carrying mites, should always be monitored. The following techniques have been selected as they are relatively easy to carry out, each has advantages and disadvantages.

Drone brood uncapping – a quick method

Examining drone brood is a good technique because mites are clearly visible on the white drone pupae. Drone uncapping provides a more accurate picture of infestation levels during the main beekeeping season than sampling adult bees. A disadvantage is that it can significantly deplete drone availability for queen mating and can only be used during the period of time when drone brood is present in the colony. Note that colonies only produce relatively few drones compared to the numbers of workers and so repeated culling, when using this technique, will severely reduce the number of drones.

The method can be incorporated into the regular colony inspection programme as follows.

- Select an area of sealed drone brood at the purple eye stage.
- Insert the prongs of a drone uncapping fork under the brood cappings and lift out the pupae; you will be able to lift out 10-20 at once.
- Repeat this for 100 drone pupae.
- As you take out each batch of drone pupae examine them and count the number of mites seen. Note that younger Varroa mites are pale coloured. The mites move rapidly away from light, so rotate the uncapping fork to ensure that you see all mites present.

There are two ways to interpret the results. The first is a relatively crude assessment carried out during the drone rearing season, typically April to August. If one in 50 pupae sampled has mites, the infestation is light and no action is necessary. If there is one in 20 the infestation is medium and one in 10 is considered heavy. If one in 6 is infested then the colony may be at risk of collapse. In the last three examples treatment with an approved varroacide is essential.

A more accurate assessment can be calculated using a percentage value for the proportion of infested drone larvae. Accuracy improves with the number of drones sampled. The subsequent action required is shown in table 12.

TABLE 12 Action thresholds for drone brood uncapping method

Period	Proportion of infested drone pupae		
March/April to June	Less than 2% *No action*	Between 2% - 4% *Plan for controls to be used*	Over 4% *Control with effective product advisable*
June and July	Less than 3% *No action*	Between 3% – 7% *Use light control technique(s)★*	Over 7% **Severe risk treat now with optimally effective product**
August	Less than 5% *No action*	Between 5% – 10% *Use light control technique(s)★*	Over 10% **Severe risk treat now with optimally effective product**

★For light control use biotechnical methods e.g. drone uncapping or varroacides that have a relatively low efficacy that will help to keep mite numbers under control.

Natural daily mite drop

In this technique the number of mites recovered from floor debris collected on sticky inserts can be used to assess whether action is required. Sticky inserts can be made by using A4 adhesive labels stapled to an insert board, or spray-on cooking oil applied to thin card and then pinned to the board. The sticky board should cover the whole area of the floor of the hive. If the natural mite fall is being measured without the use of a mesh floor then the insert should have mesh placed at least 8mm above it to prevent bees from removing some of the mites from the insert as this will lead to an underestimation of the extent of the infestation.

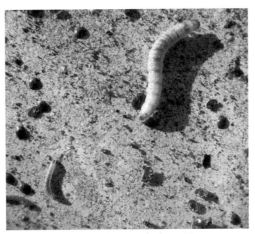

Hive floor debris including Varroa
and wax moth larvae.

The method involves:

- Using a mesh floor with the sticky board inserted underneath the mesh. The mites will fall through the mesh and stick to the insert.
- Remove the sticky inserts after at least 7 consecutive days during the summer months and count the number of mites; you may need a hand lens to see them.
- During the winter period remove the sticky inserts after 14 consecutive days and again count the number of mites.
- If there is a lot of debris which will make the mites hard to find, gather the debris into a shallow dish and mix it with methylated spirits, the mites will float to the surface whilst the wax, propolis and other debris will sink.
- Divide the number of mites counted by the number of sampling days to give the average daily mite drop figure.
- No Varroa treatment should be carried out during the sampling period, unless you want to assess the effects of a treatment.

TABLE 13 Action thresholds using natural daily mite drop

Period	Average natural mite drop per day		
January to March	Less than 2 *No action*	Between 2 and 7 *Plan control for the coming season*	Over 7 *Control with effective product advisable*
April to June	Less than 1 *No action*	Between 1 and 7 *Light control*	Over 8 **Severe risk treat now**
July and August	Less than 2 *No action*	Between 2 and 8 *Light control*	Over 8 **Severe risk treat now**
September to December	Less than 6 *No action*	Between 6 and 8 *Light control*	Over 8 **Severe risk treat now**

In the British Isles colony collapse is very likely to happen before the end of the season if the average daily mite drop for a normal colony exceeds the following values.

TABLE 14 Colony collapse thresholds

Date	Average daily mite drop value
Winter / spring	0.5
May	6
June	10
July	16
August	33
September	20

Data from the UK National Bee Unit (2009)

Estimation of the Varroa population

The UK National Bee Unit (NBU) has a Varroa calculator which can be accessed free of charge on line and used to estimate the Varroa population in your colonies. It gives levels of risk and urgency of treatment. As these levels may change in future years as a result of experience and research it is advisable to compare your results with those of the Varroa calculator each season.

Post treatment monitoring

Integrated Pest Management also recommends that you monitor the effects of treatment. Sticky inserts should be used after a treatment application to determine if the treatment has reduced the mite population. There are other monitoring methods and information about these can be obtained from the National Bee Unit and on other internet websites of national bee health authorities.

BIOTECHNICAL CONTROL TECHNIQUES FOR VARROA

Drone Trapping

Drone trapping is a biotechnical technique that can be used from mid March to mid July, during the period when drones are being raised naturally in the colonies.

Research has shown that drone larvae secrete semiochemicals (substances produced by an animal and used in communication). These include methyl palmitate, methyl linoleate and ethyl palmitate. Methyl palmitate is a pheromone of honey bees which induces workers to cap cells containing larvae. The quantities of these semiochemicals increase before the drone cells are capped and they attract female Varroa mites. It is known that pheromones found in royal jelly and brood food are also attractive to female Varroa mites.

The principle of the drone trapping technique is to stimulate the colony to produce a large number of drone larvae to attract the Varroa mites to them so that after the cells are capped the mites are isolated within them and can be removed and destroyed. Varroa mites cannot survive outside the colony.

How to use the technique

The two simplest methods are, either add 1 or 2 super combs to the brood box in spring, and allow the bees to draw natural drone comb from the bottom bars of the supers, or place a brood frame fitted with a sheet of drone foundation in the brood chamber.

1. Mark the tops of the frames with coloured drawing pins to identify them.
2. Keep a record of the dates when the drone comb was inserted

into the colony and when the cells were capped. Put a note in your diary to remind you remove the frames well before the adult drones emerge.

3. Do not allow the drones to emerge as this will defeat the object of the exercise and release additional mites into the colony.

4. When there is sealed drone brood on these selected combs, around 9 days after the eggs were laid, they should be removed and destroyed or, if you wish to reuse the comb, the cells should be uncapped and the pupae washed out.

5. Dry the comb and replace the frame in the colony to repeat the exercise.

6. Only use the technique 2–3 times per season, otherwise the colony will suffer because of the diversion of resources to rearing drone larvae.

Advantages of the technique

- It is easy to use.
- No special apparatus is required.
- It is well tolerated by the colony.
- No varroacide is used.

Disadvantages of the technique

- It is time consuming, requiring inspection of the colonies and good record keeping.
- It is of limited effectiveness as it only reduces the total number of mites in the colony as some mites will remain on the adult worker bees.
- It can deplete the strength of the colony because of the amount of resource being devoted to rearing drone brood rather than rearing worker brood.

Queen Trapping

This technique is time consuming. It involves the confinement of the queen inside worker comb cages which are faced with a queen excluder. The worker bees can gain access to the queen but she is confined on the comb. The queen is moved through a succession of three combs on each of which she spends 9 days. After each 9 days the queen is moved to a new comb cage and the Varroa mites are attracted into the cells containing unsealed larvae in the cage on the previous comb. The combs with the sealed cells are left for a further 9 days and then removed and destroyed.

Advantages of the technique

- It can be very effective.
- No varroacide is used.
- More worker bees are recruited to foraging because the queen is laying fewer eggs and they have less brood to look after.

Disadvantages of the technique

- It is time consuming.
- It can weaken or even harm a colony if used at the wrong time of the year, e.g. late summer.
- It requires good beekeeping skills.
- Frequent manipulating of the queen increases the chance of injuring her.

Using the artificial swarm technique to control Varroa

The artificial swarm technique can be used as part of an IPM programme. It is best used in the swarming season approximately mid March to mid July. It is a biotechnical method to help reduce the numbers of Varroa mites in a colony.

The artificial swarm technique in Varroa control uses the natural tendency of bees to swarm, and separates the flying bees from the majority of the mites that are present in the brood nest. In addition, as there is a period of several weeks between the emergence of the last bees from eggs laid by the queen of the original colony, and the time when new brood is being laid as a result of a new queen, the potential increase in the number of Varroa mites in the original colony is significantly reduced.

Any mites present will be on the exterior of the bee's bodies and therefore much more vulnerable to treatments such as Apistan® / Bayvoral® and oxalic acid. Its limitations are that it requires replacement queens, and precautions need to be taken to prevent the artificial swarm from absconding. The technique involves the simultaneous management of two colonies; the parent colony is A, and the artificial swarm is B.

Establishment of colony A

The parent colony **A** is monitored until the preparation signs for queen cells are seen. The queen is then caged and removed from the colony. The colony is then placed several metres away from its original site in the apiary.

Establishment of colony B

The artificial swarm **B** is created by placing the queen from colony **A**, liberated from her cage, onto a frame of drawn brood comb in a new brood box on the original site. The rest of the brood box is filled, either with drawn comb, or foundation. A queen excluder is inserted below the brood box and above the floor to prevent the queen and a swarm from absconding.

The flying bees from the parent colony **A** will relocate to their original site and form the artificial swarm **B**, and immediately start to draw the foundation (if drawn comb has not been provided). This colony can be fed weak sugar syrup to provide energy for drawing out new comb. Care should be taken to prevent robbing. This can be done by feeding the bees overnight, being careful not to spill syrup over the outside of the hive and by reducing the size of the entrance.

The Management of colony A

The parent colony **A**, minus its queen, now sets about raising the existing queen cells. After one week, remove all except one unsealed queen cell in which you can see a healthy larva. Allow this cell to develop, and when it is sealed, place a queen cell protector cage over it. This will confine the emerging queen, but enable the workers to feed and look after her. The virgin queen must not be allowed to fly and mate so she cannot lay eggs, because a key feature of this control method is an extended brood-less period in the brood box.

After about three weeks or when all the brood in colony **A** has hatched transfer two combs of unsealed brood from artificial swarm **B** into colony **A**. These will act as bait combs for the Varroa mites. When all the cells of the brood on these combs are capped, remove the frames from the colony, destroy all the larvae and with them the Varroa mites which have been attracted into the bait combs. Then replace the cleaned combs back into colony **B**.

After this you have a number of options with colony **A**:

Option 1

Remove the virgin queen and introduce a mated and tested laying queen, then maintain it as a separate colony.

Option 2

Remove the virgin queen and unite the colony with the artificial swarm B that includes the original queen.

Option 3

Remove the virgin queen from colony **A**, re-queen colony **A** and when the queen is established and laying eggs, unite with colony **B**, having de-queened colony **B**.

The Management of colony B

Leave the queen excluder in place for about a week after which the queen should be settled and laying in the newly drawn comb. Once the bait combs have been removed from colony **A** decisions can be made on the fate of colony **B**.

Option 1

It can continue to be developed as a separate colony.

Option 2

It can be de-queened and united with the newly re-queened colony **A**.

Option 3

If the queen has special qualities and you want to keep her, you can create a three or five frame nucleus colony for her; the rest of the frames of brood and surplus bees can be united with colony **A**.

If colony B is to be developed into a full colony, further Varroa control can be achieved by treating the colony with Bayvoral® / Apistan® or Apiguard®.

A modification is to carry out the artificial swarming on the original site, by placing the parent colony **A** above colony **B**, with its entrance facing in the opposite direction to that of the artificial swarm **B**. The colonies can be kept separated by a Snelgrove Board. Then choose and carry out your preferred option.

Shook swarm technique

The shook swarm technique can also be used to reduce the population of Varroa in a colony. See relevant section in 9–Useful Techniques.

THE USE OF PYRETHROIDS IN THE CONTROL OF VARROA

Pyrethroid is the general term for a class of chemicals that are effective against a wide range of pest species including insects, termites and mites. They were developed as a result of research carried out on extracts (pyrethrins) of the Pyrethrum flower (*Chrysanthemum spp.*). The active substance in Bayvoral® is flumethrin and in Apistan® it is tau–fluvalinate; these are synthetic pyrethroids. Both products are authorised and licensed for use as varroacides in the British Isles. These authorisations require a comprehensive assessment of data to show that the product meets the statutory levels of safety for bees, the consumer and the environment, and efficacy (effectiveness) against Varroa. Pyrethroids are highly effective >95% and can be used even during a nectar flow.

Development of mite resistance to pyrethroids

Resistance is the ability of an organism to tolerate toxic doses of a substance that would be lethal to the majority of individuals in a normal population of the species.

The challenge of a pyrethroid to a Varroa mite is one of life or death and if it dies, and therefore does not breed, it will fail to pass on its genetic characteristics. If the dose of the substance received is insufficient to kill it, or the mite has already developed resistance to it and its reproductive ability is unimpaired, it can continue to survive and replicate itself. Such survival may be due to particular physiological differences between mites and these may be genetic and heritable. Gradually the resistant mites begin to dominate the mite population, especially if the beekeeper continues to use the pyrethroids.

Products for the control of Varroa mites must be tailor-made. Do not be tempted to use homemade treatments from products containing pyrethroids designed to control other pests as this may promote the development of resistant strains of Varroa mites. Similarly, incorrect application or inadequate usage rates of pyrethroids will also accelerate the development of resistant strains.

Choosing the correct time of application

In general it is recognised that pyrethroids are most effective in controlling Varroa mite populations between August and October after the honey supers have been removed and there is less brood, in particular drone brood, in the colony. The treatment can be applied at other times especially if your own monitoring and the Varroa population models prepared by the National Bee Unit recommend that treatment be given before the end of the beekeeping season.

The actual time of application will vary in different parts of the country and between different beekeepers, especially those whose bees work later forage such as ling heather. Leaving the treatment too late increases the risk that Varroa and associated viruses will damage the future prospects of the colony by weakening the development of winter bees. These bees are physiologically adapted to overwinter and to start brood rearing early in the new season.

The conditions of approval for the use of Bayvoral® and Apistan® strips do not require a withdrawal period during a nectar flow or honey production. However, they should not be used during the production of cut comb and section honey. Supers containing combs being worked to produce comb honey should be cleared of bees and removed from the colonies to be treated and either placed on other colonies, or kept in a bee proof place before being replaced on the colony at the end of the treatment period.

Buying and using the products

Before you buy your products always check the following points:
- That the product is within its Use By or Expiry Date, especially check the date of low price offers before you buy from a supplier or accept product from another beekeeper.
- That the product has been stored at the correct temperature i.e. below 25°C.
- That the foil packages are still sealed at the time of receipt.

You are required by law to read the product label and the information on the packaging, and to follow the instructions for the safe use of the product and safe disposal of any residues, unused product and packaging. These precautions are to protect you and the consumer of your bee products as well as the bees. Read the product information before you go into the apiary, firstly to remind yourself about what is required, and secondly to check if the instructions for use have changed since you last used the product. This procedure should be followed for all chemicals and treatments used in beekeeping.

Correct application techniques

The information found on the product packaging describes the precautions that you should take when handling the plastic strips.
- Always handle the strips by their edges or by the top of the strip and not by the strip faces.
- Handle the strips with gloves and blunt-nose pliers to help you place them down between the combs and, perhaps more importantly, to get a grip on the strips when you remove them.
- Place the correct number of strips between the combs in the centre of the brood rearing area. The number of strips depends on the product you use and the size of the colony you are treating.

TABLE 15 Numbers of strips of Bayvoral® & Apistan® required for different colony sizes

Colony size	No. of Bayvoral® strips	No. of Apistan® strips
Nuclei, young colonies, newly collected swarms	2	1
Normal single brood colonies	4	2
Large colonies where the brood is in more than one brood box	4 per brood box	1 per 5 frames of brood

- If you have to insert or remove strips in poor weather, or at the end of a nectar flow when the bees may resent you opening up the colony, spray clean water over the tops of the frames using a hand held garden water spray to keep the bees below the tops of the frames. Also spraying the entrance discourages bees from leaving the colony to investigate.

- Record the date of application and the date when the strips should be removed. Do not leave the strips in place for more than the specified time, usually 6-8 weeks.

- Take a plastic bag into the apiary when you remove the strips, place the used strips in the bag, and wrap it in paper before putting it into your domestic refuse bin for collection and disposal.

A strategy for the continuing use of pyrethroids

It is essential that as beekeepers we aim to maintain the option of using pyrethroids as part of an Integrated Bee Health Management approach.

There are five key steps to help reduce the development of resistant mites.

- Use an Integrated Bee Health Management approach.
- Only use products containing pyrethroids that have been formulated and authorised for use as varroacides in the British Isles.

- Keep the treatment periods short and infrequent, and follow the manufacturer's directions. Keep the dosage as high as recommended in the instructions. The Varroa mite must be subjected to a high dose of the active ingredient in order to keep any resistance functionally recessive.
- If resistance to the pyrethroids is suspected, carry out a test to detect resistance and reduce the selection pressure by altering the method(s) of treatment to include the use of efficient biomechanical methods and / or varroacides that have a different mode of action, e.g. thymol-based products.
- Rotate the use of authorised, unrelated varroacide products. For example pyrethroids one year, and Apiguard® the next and a treatment of oxalic acid in winter each year to further reduce any mites present on the bodies of the bees.

Testing Varroa for resistance to pyrethroids

A beekeeper using pyrethroids should test periodically for any signs of resistance. It is likely that the first time the beekeeper considers the existence of resistant mites in colonies is when they collapse after treatment with pyrethroids has failed to control Varroa. The original test was developed with small Apistan® strips designed for use in package bees; however these strips are no longer available. The strips used in full sized colonies contain about 8 times the amount of the active substance tau-fluvalinate in the smaller strips.

The following test can be used to check for resistance.

1. Use a pair of forceps to hold the Apistan® strip, cut it into 8 equal size pieces by cutting it in half up the length, and cut each strip into four. This provides the same dose as in the package bee strips.
2. Cut a piece of thick card and cut out a window in it over which the piece of Apistan® strip is stapled. Try to minimise finger contact with the strip pieces. Place the card in a 500ml jar ensuring that the strip is not touching the glass.
3. Prepare a 2-3 mm light metal or plastic mesh cover for the jar.
4. Shake bees from 1-2 combs of a colony into an upturned roof.

Take care not to shake the queen into the roof. Scoop about 280ml (½ a pint) of these bees (about 150) and place them in the jar. Return any remaining bees in the roof to the hive.

5. Place a sugar cube in the jar. Cover the jar with the mesh lid and store it upturned in the dark at room temperature.

6. After 24 hours hit the upturned jar with the palm of the hand over a sheet of white paper three times. Count the dislodged dead mites.

7. Place the upturned jar over another sheet of white paper, place it in a freezer and leave until the bees are dead (1-4 hours). Count the remaining mites which have fallen off the bees. These are mites that were alive until the bees were frozen and therefore not knocked down by the pyrethroid treatment.

8. If the total number of mites per jar is below 5 there is no resistance.

9. Otherwise calculate the percentage mite kill caused by the pyrethroid. Less than 50% indicates that you may have pyrethroid resistant mites.

10. This method gives only an indication of resistance and confirmatory tests are advisable.

Techniques continually evolve and readers are encouraged to check with their bee health authorities for recent advances in resistance testing.

OTHER TECHNIQUES TO CONTROL VARROA

The use of Apiguard®

Apiguard® is a product authorised for the control of Varroa. The active ingredient is a terpene called thymol contained in a gel matrix (25% active ingredient), it provides a vapour that is slowly released into the hive. Follow the manufacturer's instructions and apply two 50g treatment packs, one initially and the second 10-15 days later. The product is normally applied in the spring before a nectar flow begins, or late summer after the honey harvest has been removed from the colony. Apiguard® is 90-95% effective in the optimum conditions of an ambient temperature of more than 15°C and active bees. This is because for thymol to be effective, the ambient temperature needs to be high enough for long enough to release the thymol vapour. To ensure a vapour concentration high enough to be effective place a board under each Varroa mesh floor and close the vents in the crown boards.

The use of Api Life Var®

This product has been approved for use as a varroacide in the UK. It contains thymol, eucalyptus oil, menthol and camphor carried in an inert strip which is applied by breaking it into 4 pieces that are placed on the top bars of the brood box after the removal of the honey crop. In the UK there is no withdrawal period, no reported resistance and the manufacturer recommends two applications, each using 2 strips, applied 14 days apart. The manufacturer's instructions should be carefully read prior to use and then followed during the product application in order to ensure optimum effectiveness of the product. As with all thymol-based products it is most effective when applied during warm weather.

The use of oxalic acid

Oxalic acid solution is used to treat Varroa during the winter and brood-less periods. It works by damaging the claspers on the proboscis of the Varroa mite preventing it from sucking haemolymph from the host bee. It is not authorised in the UK as a Varroa treatment but its use by beekeepers is tolerated by the regulatory authorities at the time of writing. This situation may change and the beekeeper is advised to check the latest position.

Beekeepers wishing to use oxalic acid are recommended to buy a proprietary product formulated with sugar. One product for example contains a 6% solution of oxalic acid in a 30% sugar solution. Another comes in the form of a 3% solution of oxalic acid supplied with sugar to make up the desired concentration. Remember to read the product instructions and follow them and when you have finished the treatment update your hive treatment records stating product used, overall dose and the date on which the solution was applied. It is advisable to give only one treatment per year. The oxalic acid solution has a short shelf life once it is mixed with sugar because the hydroxymethylfurfuraldehyde (HMF) level in the treating solution rises and can become toxic to bees. It is recommended that any unused solution is safely disposed of down the drain.

Oxalic acid can be applied by trickling or spraying. Trickling is the easiest in terms of efficacy and application and is said to be around 90% effective. A syringe without a needle is used to gently trickle 5ml of a lukewarm solution over each seam of bees between two frames in situ. Ideally choose a bright and warm day when the bees are loosely clustered and, by moving through the cluster, they will distribute the chemical onto the mites.

Oxalic acid is lethal to brood so should only be applied during brood-less periods usually this will be in winter. Treatment during the time when there is brood is ineffective because any mites present inside the brood cells will be a residue population and will not be affected by oxalic acid. Oxalic acid cannot be fed directly in sugar syrup when feeding your colonies, for example in preparation for winter, as the bees will not consume the treated syrup. Oxalic acid can also be applied by using evaporators to heat oxalic acid crystals inside the hive so that it sublimates onto the bees. This technique is not recommended as it significantly increases the risk of inhaling the sublimate for both beekeepers and bees and is no more effective than the trickling method described above.

The use of oxalic acid will probably increase because its efficacy does not depend on the temperature needed for evaporation as is the case with products containing thymol.

The use of icing sugar

This technique can be used any time during the summer months to help reduce the numbers of Varroa mites in a colony, and should be used in conjunction with mesh floors. Ideally granulated sugar that has been powdered in a food processor should be used, but bought icing sugar is adequate. Between 75 and 100g is sprinkled over a travelling screen placed above the brood box. A soft paint brush is then used to push the sugar through the mesh onto the frames of brood and bees. It works because honey bees will groom themselves to remove the icing sugar, this dislodges the mites which fall through the mesh and are unable to crawl back up onto the bees. The dislodged mites appear on the Varroa insert board 10-15 minutes after the sugar application.

Beware of using unauthorised substances

The internet has enabled beekeepers to become aware of products and substances used by beekeepers in other countries to control Varroa; some can be bought over the internet. Beekeepers should always check on the legal status of these products before attempting to buy and import them for use in their apiaries. Criminal proceedings may be taken by the regulatory authorities against beekeepers who use them. However, if you chose to take this approach you must be aware of the potentially harmful effects on both bees and the beekeepers applying them, and the possibility that the products will leave residues or contaminate bee products such as honey, wax and propolis.

Any product or chemical substance applied to bee colonies or to beekeeping equipment has the potential to leave residues in bee products (honey and wax) as well as potentially harming adult or brood honey bees. Also they may harm beekeepers. The following principles should always be followed to minimise risks to your bees and yourself:

- Only use products authorised by the appropriate regulatory authority, and with a proven track record. Do not use unauthorised alternatives or homemade products that may lack reliable residue and effectiveness data.
- Always read and follow the directions on the label supplied with all authorised products.
- Use personal protective clothing, gloves, face and eye protection, as described on the label or product information.
- Never treat colonies immediately before or during a nectar flow, or while supers are on the hive, unless the label directions of an authorised product specifically permit it.
- Buy your products from a reputable supplier.

Some products and substances have no formal authorisation because there has been no regulatory assessment of the efficacy or safety to honey bees or beekeepers, the environment, and bee products. Beekeepers who choose to use unauthorised products should be aware of the risks involved.

The regulation and control of substances that may be used in the control of Varroa is constantly evolving and changing in the light of research, experience and advances in knowledge and understanding.

If in doubt seek advice from government bee health agencies or the product supplier. Beekeeping Associations can also provide information on the current status of various substances used in beekeeping.

Biological control of Varroa

This has been the subject of extensive research and at the time of writing there is still much work to be done to translate the research findings into a biological control product that beekeepers can use successfully to control Varroa. To date various microbial species have been identified as being of potential use in biological control.

The timing of Varroa management techniques

A number of techniques already described can be used in combination to control the Varroa population in the colony. The following table shows the time of year when these techniques can be used. Timing will vary depending on the geographical location of the apiary, the levels of Varroa mites present and other factors such as the weather and your management programme.

TABLE 16 Options For Varroa Management Techniques

	Jan	Feb	Mar	Apr	May	Jun	Jul	Aug	Sep	Oct	Nov	Dec
Open mesh floor	■	■	■	■	■	■	■	■	■	■	■	■
Drone brood removal				■	■	■	■					
Drone comb trapping				■	■	■	■					
Apiguard®								■	■	■		
Apistan® / Bayvoral®			■	■	■			■	■	■		
Oxalic acid	■											■
Icing sugar dusting					■	■	■	■				

■ Period during which technique can be used

SECTION 4

ORGANISMS AFFECTING
HONEY BEE HEALTH

ORGANISMS THAT DAMAGE HONEY BEE COLONIES IN THE BRITISH ISLES

Wax moth — Greater (Galleria mellonella) and Lesser (Achroia grisella)

The larvae of these moths have the ability to digest wax and their preferred food is the old larval and pupal skins found in brood comb and in honey supers in which patches of brood have been raised, honey stored and subsequently extracted. The larvae tunnel through the comb just below the cappings and produce silken tunnels to which their faeces and bits of wax become attached. The whole comb can be destroyed and the honey in affected supers is contaminated. At the time of pupation the larvae of the Greater Wax Moth become gregarious, they line up in rows and excavate hollows in the woodwork of the hives, especially the frames, and any recesses in which they can pupate and it is very difficult for the bees to remove them. If the frames are badly damaged by these excavations, even when cleaned and fitted with new wax foundation, bees will often avoid them. Good practice is to discard damaged frames. The Lesser Wax Moth does not have this habit and usually pupates in the comb.

The adult moths of both species usually gain access to the active colony only if it is weakened by Varroa or other diseases, although some of the early designs of Varroa insert trays enabled the moths to lay eggs, and the larvae to develop and access the colony without the bees being able to remove them. Current Varroa floor designs have eliminated this fault. Colonies are accessed at night by female moths soon after they have emerged and mated. Keeping healthy colonies with narrow entrances for the bees to defend reduces the potential for the moths to gain access. It is good practice when removing the roof and crown board to check them for the presence of adult moths and remove them.

The majority of damage caused by the Lesser Wax Moth occurs in boxes containing drawn comb that is being stored overwinter. Either the frames already contained eggs, larvae, pupae, or moths before they were stored, or the wax moths subsequently gained access to the comb and laid eggs.

Wax moths are a real problem in stored comb and the only sure way of keeping the combs from attack is to check for moths in the comb; if you believe them to be free of wax moth place them on a crown board with its holes covered with a piece of gauze. Then seal the areas between the boxes with tape and place another crown board on top with its holes also covered with a piece of gauze.

If there is a possibility that your comb may contain wax moth in any form put it in sealed boxes and fumigate with the vapours from a dish of 80% ethanoic acid (acetic acid) placed on the top of the frames inside the box; the vapours are heavier than air and will spread down over the frames. Acetic acid vapours are corrosive to metal parts of the hive and this can be prevented by covering them with Vaseline®. Take care when handling and using ethanoic acid. The boxes containing the combs should not be placed directly on concrete as the acid vapours will corrode it, so place them on a sheet of plastic. The combs should be fumigated for a minimum of 7–10 days and should be well aired before use.

For many years beekeepers used paradichlorobenzene (E680) crystals and moth balls containing naphthalene. The Honey Regulations (England) (2003) and equivalent legislation in the devolved administrations require that no trace of either of these substances is present in honey for consumption in the British Isles and other EU Member States, so beekeepers should not use them to deter wax moth. These chemicals will contaminate the wax in the comb as well as the honey placed in it subsequently.

A biological control technique can be employed using a suspension of the bacterium *Bacillus thuringiensis*, (a specific parasite of lepidopteran larvae). This is marketed as CERTAN® or B401. The preparation is mixed with clean water and sprayed onto the comb before storage. The product can also be used when making up new frames; it is sprayed onto the wax foundation and allowed to dry before the combs are inserted into the hive.

Another method of killing wax moth present in comb is to place combs in the deep freeze at −18°C for 24 hours. Care should be taken when handling the cold combs because they become very brittle at low temperatures. After removal from the freezer store the combs in supers with the edges taped to prevent wax moth access.

Spray application of a solution of Certan® product on
foundation to control wax moth.

Freezing is also very useful to prevent deterioration in store of cut
comb due to the presence of wax moth eggs, larvae, pupae or any adult
moths prior to sale.

Wasps

There are three species of wasps that are relevant to beekeeping in the
British Isles; they are *Vespula vulgaris, V. germanica,* and *Dolichovespula
saxonica.*

At the end of the summer wasps can become a nuisance by robbing
and killing weak colonies. There are a few simple techniques that
beekeepers can use to help bees to resist wasp attacks.

- Reduce the entrance to one or two bee body widths to help
  the guard bees defend the entrance.
- Use open mesh floors together with narrow entrances. If
  wasps fail to gain easy access to the entrance they will try to
  get through the mesh floor and in my experience this distracts
  them from the entrance. In addition, the movement of wasps
  under the mesh floor attracts other wasps and they are also
  diverted away from the entrance.
- Ensure there are no holes in the hive parts through which
  wasps could gain access.
- Do not spill sugar syrup when feeding the bees, clear it up if
  any is spilt.
- Do not leave comb lying around in the apiary.
- Set up wasp traps, e.g. a honey jar with a 10mm hole pierced
  in the lid, half filled with a jam/water mixture. Do not use

honey as this will attract bees into the trap.

- From late autumn onwards look under the roof and cover board for over-wintering queen wasps and decide whether you want to kill them.

Hornets

Hornets of the species *Vespa crabro* are now seen by an increasing number of beekeepers in the British Isles, possibly because of an extension of their range as a result of climate change. They can attack and destroy a colony, but they are more frequently seen flying through an apiary taking honey bees in the air or from the alighting board, and they will fly through swarms. As with wasps, the use of small entrances will usually prevent the hornets gaining access to the hive. If hornets become a real problem you may have to resort to finding the nest and destroying it, wear protective clothing for this. If you cannot locate the hornet's nest you will need to relocate the hives to another apiary.

Mice and shrews

In the autumn shrews and especially mice, can gain access to the colonies and will destroy comb and bees. The presence of mice in a colony is typically revealed by pieces of wax comb on the alighting board.

The tradition is to use mouse guards which are pieces of sheet metal placed across the entrance to the hive, with holes or slots through which the bees leave or access the hive. However, a disadvantage in their use is that honey bees, foraging for the all important early pollen in late winter and early spring, often lose their pollen loads as they crawl back though the holes or slots which dislodge the pollen. Colonies are particularly vulnerable to mice and shrews if the entrances are near the ground. Raise the hives onto hive stands and use very narrow entrances (both vertical and horizontal), in conjunction with open mesh floors to deter mice and shrews without using mouse guards,

Woodpeckers

The Green Woodpecker (*Picus viridis*) can damage or even destroy hive parts and eat brood and bees causing the demise of the colony. The parent will also train its young in the art of finding a good meal in a beehive. The most effective way of preventing woodpecker damage is to place a cage, made from chicken wire, over the hive with sufficient distance between the mesh and the hive parts to prevent the woodpecker's bill reaching the hive.

Badgers

Badgers (*Meles meles*) are strong animals, easily capable of overturning hives and gaining access to the brood which they relish as it is an extremely useful food for them. Prevention measures include placing hives in areas not frequented by badgers; if this is not possible, fence the apiary area with netting sunk into the ground so that badgers cannot get into the apiary.

Sheep, horses, cows and deer

Beehives are excellent animal rubbing places, if allowed access all these mammals will utilise the facility. This can dislodge the hives or even overturn them with the potential loss of bees. In some cases the bees may attack the animal in sufficient numbers to kill it. This is not a desirable outcome because of the economic loss to the stock owner, the potential colony loss and the loss of reputation for beekeepers.

Always obtain specific permission from the owner to place your hives on their land. Ascertain whether sheep, horses, cows or deer might have access to the site, before moving your hives to it. If animals are present ask the owner if you may put up a temporary post and wire fence to enclose and protect the colonies, or alternatively ask where the hives could be better sited to prevent access by stock.

OTHER ORGANISMS SOMETIMES FOUND INSIDE BEEHIVES

There are a number of other species which beekeepers may see when examining their colonies.

Death's-Head Hawk-Moth (Acherantia atropos)

This is the one of the largest hawk-moths in the UK and when entering a hive it emits a sound similar to the queen bee that has a calming effect on the bees so they do not attack the moth. The sound is produced by the moth expelling air through the proboscis emitting low and high pitched squeaks. The moth is then able to freely move around the colony consuming honey without interference.

Earwigs (Forficula auricularis)

Earwigs can be present in large numbers in colonies and can cause damage to honey combs by piercing the cappings. One method of keeping the numbers down is to frequently brush all the crevices and other hiding places, especially under the roof at the junction with the crownboard. Standing the feet of hive stands in containers with a water/disinfectant mixture prevents the earwigs from gaining access.

Ants (Formicidae)

There are several species of ant which if they gain access to a colony can cause it to become demoralised and contaminate the honey with soil. Standing the feet of hive stands in containers as above also prevents ants from gaining access.

Spiders (Arachnidae)

There are various species of spiders which will live in beehives. Their webs block ventilation and bees can become trapped and consumed by the spiders. Some spiders actively hunt bees e.g. the wolf spider and can be why virgin queens are often lost for no apparent reason. Regularly brushing the webs from crevices and removing the spiders are good preventive measures.

Bee Louse (Braula coeca)

This is a wingless insect that was once much more common than it is today. Chemicals to control Varroa also kill braula. The adult feeds on nectar and pollen (and possibly saliva) at the bee's mouth. It tunnels under the honeycomb cappings causing the honey to seep out. In order to kill the insect in the honeycomb, especially comb destined to become cut comb, freeze it for 7 hours or keep it frozen until it is used.

Beetles (Coleoptera)

There are several beetles which can be found in beehives such as *Ptinus sexpunctatus,* ladybirds and puffball beetles. Any beetles found in the hive should be carefully examined to see if their description fits that of the Small Hive Beetle.

Pollen mites

In terms of identification pollen mites are smaller than those of Varroa and require confirmation in the laboratory. Brood combs which contain large amounts of pollen can become infested with several pollen mite species that consume the pollen and leave behind dry dusty remains. The best way to prevent such infestations is to remove and destroy old combs which contain a lot of pollen.

EXOTIC PESTS

There are a number of honey bee pests that are currently not found in the British Isles. Experiences with Varroa have demonstrated the impact of exotic pests if they become established and are not successfully eradicated. There is a continual process of monitoring for such pests and contingency plans in place to deal with them. In spite of these provisions beekeepers should be vigilant and report the occurrence of any unusual organisms found in their hives to the bee health authorities. The following exotic pests are currently considered to be the greatest threat at the time of writing.

Small Hive Beetle (SHB) (Aethina tumida)

This beetle presents a significant threat to honey bee colonies. As of 2009 it has not been found in the British Isles and it is the subject of a surveillance programme designed to provide early detection of entry into this country. All beekeepers should be alert to the possibility of SHB being present in their colonies. Experience in other countries where SHB is present shows that the maintenance of strong colonies helps in the protection of the comb and defence of the brood nest. Stored supers and other boxes that contain or have contained honey will attract the beetles. The honey extraction area should be kept free of hive debris and honey supers extracted quickly.

The key identification features of SHB are:

Adult beetles

- Size 5–7mm
- Black coloured
- Club–shaped antennae

- The behavioural habit of hiding from light
- Short wing cases

Larvae

- Size 10-11mm
- Beige coloured
- Have spines on the dorsum*
- Have 3 pairs of prolegs*
- No frass or webbing on the comb*
- Active in light*

*Compared with wax moth larvae which have no spines, and produce frass and webbing in their tunnels. The wax moth larvae do not have prolegs and they move away from the light.

Eggs

- Size 1.5 x 0.25mm
  (two thirds the size of honey bee eggs)
- White coloured
- Large clusters of eggs in hive crevices or in the hive floor.

Techniques and experience in the management and control of SHB in other countries are continually evolving so check the latest advice. The techniques being developed rely on the removal of the SHB at different life stages to disrupt its life-cycle. One monitoring method uses corrugated paper trap inserts that can be placed on the floor of the hive; the beetles move inside the corrugation away from the light and can then be removed and destroyed.

If you find beetles or larvae in your colonies that fit the above descriptions, kill them by keeping them in a freezer overnight, or by putting them in 70% ethanol (e.g. methylated spirits). Once dead, pack them into a container such as a plastic tube or stiff cardboard box, ensure it is well sealed and send it to your national bee health authority for examination. Include your name, address and the location where the beetles were found giving an Ordnance Survey map reference if possible. Alternatively telephone your local bee health inspector immediately. Do not move any of your colonies or bring new ones into the affected apiary.

Asian Bee Mites (Tropilaelaps spp.)

The Asian bee mites *Tropilaelaps clareae* and *T. koenigerum* are native to Asia but have spread from their original host, the Giant honey bee *Apis dorsata* to the European honey bee. As of 2009 they had not been found in the British Isles and beekeepers and the bee health authorities are required to maintain constant vigilance.

These mite species rely on the presence of brood for feeding and can cause serious problems such as abnormal brood development and the death of both brood and adults bees, leading to colony decline and collapse. If the bees abscond they will spread the mites to other colonies. Colonies of honey bees heavily infected with *Tropilaelaps spp.* show similar damage to that caused by Varroa so in the event of their spread to the British Isles, it is important to be able to distinguish between them. Varroa mites are larger, crab-shaped and wider than they are long and they move relatively slowly. Tropilaelaps are elongated (1.0mm long x 0.6mm wide) and move rapidly across the brood combs.

As their life cycles are similar the main detection methods for finding Varroa can also be applied to Tropilaelaps. Good knockdown can be achieved using pyrethroid-based acaricides, however, in the event of you having to use them, check the product is specifically authorised for control of Tropilaelaps.

Africanised bees

These are hybrids between the African honey bee *A. m. scutellata* and various European honey bees, e.g. *A. m. ligustica* (Italian) and *A. m. carnica* (Carniolan bee). They are aggressive and will attack anyone who approaches within 30 metres of the colony and pursue them for at least 400 metres if they run away from the hive. This exaggerated response of the colony to the release of small amounts of the alarm pheromone isopentylacetate will result in large numbers of the bees erupting from the colony and joining in the attack. Small swarms of the Africanised bee can take over a European honey bee colony by invading the hive, killing the queen and establishing their own queen.

In theory it is possible that Africanised bees could reach the British Isles and evade the detection regime set up to monitor their arrival and presence. They can survive low temperatures by clustering, however they seem not to make enough stores of honey for over-wintering.

They tend to use forage to produce brood rather than honey stores and then when the forage fails they swarm and move on which is why they can spread so rapidly. In view of climate change it is conceivable that if the Africanised bee does arrive in the British Isles and eradication programmes fail, it could maintain itself in areas where there is a continuous supply of forage throughout the year, together with temperatures capable of sustaining active foraging and brood rearing throughout the winter months.

Asian Hornets

The Asian hornet *(Vespa velutina)* arrived in the Bordeaux area of France in 2004/5, probably from China in a consignment of plant pots. It has spread at the rate of 50/60 miles per year and colonies are now established in northern France and predicted to arrive in the UK within 5 years. It is similar in size to *V. crabro* and distinguishable from other hornets by its yellow legs and black thorax. Colonies can reach 5,000–15,000 individuals and beekeepers and the public are advised not to try to destroy the hornet colonies themselves but to seek help from their local authority. The colony becomes a significant honey bee predator when its size reaches a peak in September / October and the honey bee colonies are declining. The impact can be reduced by trapping hornets in specially designed traps.

If the Asian Hornet reaches the UK our honey bees may not exhibit the defence mechanisms used where Asian hornets and honey bees co-exist in other parts of the world, such as balling, organised mass defence and filling the hive entrance with propolis containing small entrance holes only big enough for a honey bee. The French bees unsuccessfully try to sting the hornets and continue their normal foraging behaviour and so are vulnerable to being taken on the wing.

Contingency plans will be established should the Asian hornet arrive in the UK and beekeepers are advised to look more closely at any hornets they see in their apiaries and to periodically check the situation with their bee health authorities.

SECTION 5

THE ESSENTIALS OF GOOD BEE HUSBANDRY
CHOOSING YOUR EQUIPMENT

Hives

Modern beekeeping practices and the development of hives with movable frames stem largely from the identification of the 'bee space'. This was recognised by the Reverend Lorenzo Lorraine Langstroth in 1851 in the USA; he observed that bees left a space of about 6mm (0.25 inches) between their combs. He proposed that if beehives could be designed such that a bee space of 6mm was left between all the separable parts (i.e. frames and boxes) the bees would not propolise them or build brace comb. This led Langstroth to refine the design of a movable frame hive, which he called the Langstroth Hive; it is the basis for all modern hive types. For more information on types of hives see Section 9.

Care in selecting and maintaining hives pays dividends in bee health management terms and a description of types of hives to suit your requirements and how to extend their useful life can be found in Section 9. Correctly fitting and matching parts enables easier manipulation of the combs and boxes. Weather-tight hives keep bees dry and enable them to control the environmental conditions around the brood nest. Hives without holes or gaps ensure that robber bees and wasps find it hard to gain access to the colony.

It is important to keep hive floors out of direct contact with the ground by using a hive stand to raise them to about 0.5-0.7m above the ground. This also reduces the strain on the beekeeper's back when lifting the boxes and carrying out colony inspections.

Personal Protective Equipment

It is essential that the beekeeper and any onlookers are protected from the possibility of bee stings. Even though a beekeeper may have kept bees for many years and been stung frequently, there is always the possibility of developing a reaction to stings and, in the worst situation, anaphylactic shock. Guidance on bee stings can be found in Section 9. The best protection is gained from a purpose-made bee suit and the purchase of one to meet your needs is strongly recommended. Various styles are available consisting of either an all in one type (veil, top and trousers), a boiler suit or overall with a separate veil, or a top with an integral veil and separate trousers. Beekeeping in hot weather can be stressful with sweating being a particular problem, so when selecting your clothing do not necessarily buy the thickest material. Bees get caught up in wool and become angry so woollen outer wear should not be used.

White or lighter colours such as very pale blue or khaki are accepted by bees. White will reflect heat and help to keep the wearer cool. Select clothing with strong zip fasteners, and choose designs where there are Velcro strips or elastic cuffs on the sleeves and the trouser bottoms. There are several designs of veils available and, even if you buy a bee suit with an integral veil, it is worth having a separate hat and ring veil to put on for a quick visit to your apiary, or if you are called out to deal with a swarm. The face is the most important area to protect from stings, especially the area around the temples and throat.

It is good practice to wear gloves at least until you develop more experience of working with bees. You may then consider taking them off to do delicate manipulations such as marking and clipping queens. Gloves can be made from materials including cowhide, goat-skin, kid leather and plastochrome at a range of prices. It is also a good idea to have a pair of sturdy gloves for use when moving hives. For working the bees there is much to recommend the use of disposable household gloves because gloves quickly get spoiled by propolis. Rubber or latex gloves are relatively cheap, the manual dexterity is good and perhaps most importantly they can be discarded when soiled thus reducing the potential for transfer of diseases.

Lastly it is good practice to protect the feet and legs and the best way to achieve this is to wear Wellington boots, especially those that have a good fitting around the calf. Tuck the bee suit trouser legs inside the boots. Often the top of the boot is at the same height as the hive entrance and bees are attracted to the scent of feet!

A selection of useful equipment.

APIARY SELECTION AND LAYOUT

Choosing your Apiary location

When establishing a new apiary, there are some serious considerations to be made in relation to humans, companion animals and livestock. Not everyone shares our enthusiasm for bees, and it is best to avoid keeping them where they may be a nuisance to family and neighbours, especially people who are sensitive to bee stings and could die as a result of anaphylactic shock. Do not keep bees close to public paths and rights of way, including bridleways. If you have hives in your garden ensure that foraging bees have to fly up and over hedges or fences to avoid situations where they will fly to and from the hives in direct lines at heights of less than 2 metres where they are more likely to come into contact with humans and companion animals.

Apiaries away from home

If you intend to keep your bees in an out-apiary (a location away from your garden) you should always ask the landowner for permission to position your colonies in a secure site. If you want to keep bees on an allotment it is important to check with the local authority whether there are any restrictions. When you have located a potentially suitable site, survey the surrounding area to assess if there is likely to be sufficient forage to sustain the number of colonies you intend to keep. Of key importance is a continual source of pollen throughout the months when the colonies are rearing young, i.e. from February through to late September or even October. Colonies only store about 1kg of pollen during the summer which is about one week's worth of pollen requirements in the comb, so any fall in pollen availability could impact on the health and development of brood emerging 2-3 weeks after the pollen shortage. When carrying out your survey remember that if conditions are right your bees will visit flowers within a 3-5 km (2-3 mile) radius of their hives.

Also consider whether you will be able to access the site in all weathers, and remember that the hives could be heavy with honey at the end of the season and difficult to handle, especially if the area around them is muddy and the site some distance from your vehicle. Lastly you should consider the security of the site. If the colonies can be seen from roads or footpaths there is an increased risk of theft or vandalism.

Flooding

In recent years there has been an increasing frequency of extremely heavy rainfall events leading to flash flooding. Beekeepers should think carefully before placing colonies in locations where flooding is likely, or where there are streams draining water from other areas. Generally this means avoiding flood meadows, low-lying areas near rivers and sites that are always muddy. Check the ground conditions at the time of hive location, and if in doubt, raise the stands on blocks to ensure they will remain stable, especially when they contain all those additional pounds of honey waiting to be harvested.

INSPECTIONS

A quick inspection

Start by examining the hive entrances even when the bees are not flying and record what you see. The following table describes how to interpret this.

TABLE 17 External indicators of a colony with health problems

If you see	*What can it mean?*
Dead larvae being carried out of the hive	Disease
Dead larvae on the alighting board may look as if they have been sucked dry	Possible starvation
Mummified larvae at the entrance or on the alighting board	Chalkbrood
A pile of dead bees at the entrance or on the alighting board, all at the same stage of decomposition in the pile	Suspect poisoning or starvation
A pile of dying bees as above, with dead bees at the bottom of the pile, and those at the top still moving	A paralysis disease caused by a virus
Bees unable to fly, staggering around, K-wing displacement or bees with bright, black, shiny bloated abdomens	A paralysis disease caused by a virus
Faecal spotting around the entrance and on the front face of the hive	Dysentery or Nosema
Foul smell coming from the hive	Foul brood disease or dead colony
Bees fighting at the entrance	Robbing
Large pieces of wax at the entrance	Mice in the hive

TABLE 18 External indicators of a healthy colony

If you see	*What can it mean?*
Pollen being taken into the hive	Usually indicates a healthy colony with young brood
Many bees flying around the entrance and no fighting, bees moving up and down in their flight	Young worker bees on orientation flights
Many bees issuing from the hive in a swirling and ascending mass	The colony is swarming
Many drones flying	Normal for a colony, especially around 2-4pm from late April until August
Bees fanning at the entrance	Colony is too hot or bees are evaporating water from nectar
Bees fanning, exposing Nasonov glands	The queen is on a mating flight, or a cast has issued but the virgin queen has not emerged, has been lost; or a swarm is occupying a previously unoccupied hive
Small pieces of wax at the entrance	Bees uncapping stores

In early spring, and when the ambient air temperature is below 14°C, it is too cold to risk opening up the brood nest and examining the combs; the brood will soon become chilled and die if exposed to low temperatures. However, it is possible to make a quick preliminary inspection to see if all is well. Carefully remove the roof, and look through the holes in the crown board, note the smell coming from inside the hive. If it smells yeasty and musty, check if the colony is alive. If it is dead close the entrance as soon as possible and remove the colony from the apiary. This is important for preventing robbing and the spread of disease. You can examine it later to determine the cause of death. Often it will be isolation starvation when, even though there are adequate stores in the frames, long periods of cold weather can cause the cluster to stay in one position and consume all the stores in the comb immediately under it. This happens when the temperature is too cold to allow the cluster to move onto fresh areas of comb and food. There may be other underlying causes that have reduced the number of bees in the first place, so look for signs of disease and other conditions.

It is useful to have a torch to light up the frames during a quick inspection. Do not use smoke and do not jar or kick the hive. If the bees are not clustering under the holes in the crown board carefully prise it off and move it to one side. Use cover cloths that you can unroll as you inspect each section of frames; two cloths enable you to expose the top of frames one at a time, keeping the others covered. Without removing the frames, look for sealed brood and the types of cappings. Flat cappings indicate that the sealed brood will produce workers, whilst dome-shaped cappings, at this time of year, indicate that the queen may be a drone-layer. This may be because she has run out of sperm or has limited amounts of it, she will lay unfertilised eggs and you will see drone brood cells on the face of the comb, mixed with normal worker brood. Record this, and when you make your first full inspection investigate it further. After your inspection, which should be a quick as possible, gently replace the cover board and the roof.

A full inspection

This can be carried out when the weather is warmer and the bees are flying.

1. First plan what you are going to look for, the actions you intend to take, and how to record your observations.
2. Assemble the equipment you will need for your inspection including the water spray.
3. Light your smoker and wear your personal protective equipment.
4. Look closely at the exterior of the hive for brown stains or smears of bee excreta, especially around the hive entrance; these may be a sign of disease.
5. Note whether there are bees crawling aimlessly in front of the hive, this is another indication of disease. If you are unsure of your diagnosis or how to treat it ask your beekeeping association or local bee inspector for help and advice.
6. Gently smoke the entrance and wait a minute before starting your inspection.
7. Stand at one side or behind the colony, carefully remove the roof, and check for wax moths and wasps, note which is the front, upturn it and set it to one side. If you are not using supers, gently smoke the entrance and across the tops of the frames in the brood box. Insert your hive tool between the

bottom of the brood box and the top of the floorboard or mesh floor. This may be fixed down with propolis, so carefully ease around the outside of the brood box / floor interface separating the two parts.

8. Lift off the brood box and set it diagonally across the upturned roof. Then, either replace the floor with a clean one or scrape it clean, making sure the scrapings go into a container taken to the apiary for the purpose. Do not scrape the debris onto the ground as this can cause the spread of diseases that may be present in the hive or promote robbing and attract wasps.

9. Lift the brood box back onto the floor, orientating it in the same way as it was before you moved it. Gently smoke across the top of the crown board before carefully prising it off, inspect its underside in case the queen is there, if she is, carefully transfer her onto the frames of the brood box. Also deal with wax moths and wasps if present.

10. Remove the brood comb or dummy frame from the outside edge nearest to you; hold the comb over the brood box in case the queen is on it. If queen cells are present dislodge the bees with a brush, if there are no queen cells it is safe to shake the bees into the brood box, and inspect both sides. Follow the same procedure if you are working with a brood and a half as a unit.

11. If you have honey supers on your colony, smoke before you lift the crown board and remove all supers above the queen excluder, stacking them on top of the upturned roof in the order in which you removed them, smoke again and use your hive tool to ease off the excluder, check its underside for the queen, if she is present, gently transfer her into the brood box.

12. Proceed with your inspection of the brood frames as above.

What to look for

- ## Check the colony size

Record the number of frames with brood on them. As you remove and replace the frames take care not to roll or crush the bees. Assess the colony size and compare it with that of your last inspection.

- ## Check the colony is queen-right

The term queen-right means that a queen is present, laying eggs and producing worker brood. Do not worry if you do not see the queen, there are other ways of knowing that she is present (see below). If you do see her, check if she is marked and note the marking colour. If you see dome-shaped cappings check at your next inspection to see if the amount has increased, if it has increased it is almost certain that the queen is a drone layer. Find and kill her and unite the colony with one that is queen-right.

If you do not see the queen and the areas of dome-shaped cappings are small and irregularly spaced on the comb, look for cells containing several eggs laid on the walls of the cells. If these are present it is likely that the colony is queen-less and laying workers are present. You can check for the absence of a queen by inserting a frame of brood containing eggs from another disease-free colony. If the bees do not raise queen cells you know that the colony is queen-right. If you are carrying out this test early in the season and queen cells are raised then the colony has to be united with one that is queen-right as there may not be enough drones to ensure that virgin queens will be successfully mated.

The colony can be united with a queen-right colony using the newspaper technique, see Section 9 - Uniting colonies. If you use this technique it is advisable to cage the queen to prevent the laying workers from killing her. She can be released after 2-3 days or after the bees have eaten through the paper closure.

There is another way to deal with a queen-less colony if it occurs early in the season. If the colony is disease-free move the hive approximately 200 metres from its existing site and shake the bees onto the ground allowing them to find their way into other colonies in the apiary.

- **Check the brood is healthy**

There should be a single egg in the bottom of each cell and the larvae should be pearly-white in colour. Unhealthy brood larvae can look distorted or discoloured, and unhealthy sealed brood may have moist, dark and sunken cappings. See Section 3 – Honey Bee Diseases and Conditions. When checking the pattern of sealed brood you may see cells which are empty and not sealed. Sometimes these empty cells follow the lines where wire is present in the foundation but not always and around 5-10% of the cells in the capped brood area may be empty. These cells if closely watched will be seen to be occupied from time to time by so-called heater bees. Bees maintain the temperature of the brood either by:

1. Pressing their thoraces onto the surface of the cell cappings and then elevating their body temperature by the pumping actions of their abdomens.
2. Running actively around the surface of the comb.
3. Entering the empty cells to transfer their heat to adjacent cells containing developing larvae and pupae. These heater bees cannot sustain the elevated temperatures for long periods (typically 3–30 minutes) after which they have cooled down and they leave the cells.

- **Observe the adult bees**

Check for large numbers of dead or crawling adult bees on the ground in front of the hive. Observe and record any deformities on bees in the hive, e.g. deformed wings. Section 3 - Honey bee diseases and conditions.

- **Check there is sufficient food**

In spring there is a real risk of the colonies running out of stores and it is essential to make some estimates of the food reserves during your inspections. There are also times, even in the summer, when there is no nectar flow, or bad weather has prevented bees from foraging, and stores can become dangerously low. As you inspect the frames record your estimate of the total amount of stores present. One full British Standard deep or brood frame holds 2.27 kg of stores. If you think there is less than 5kg in total, the bees should be fed using a contact feeder containing sugar syrup at a concentration of 1kg sugar to 1.2 litres of water.

- **Check the condition of the brood combs**

There are several good health reasons for regularly changing the combs in the brood box and no comb should normally be more than 3 years old. Old comb can accumulate diseases and regular comb change is a good integrated bee health management tool in the control of a number of brood diseases. At the first full inspection of the season begin the process of moving the older combs to the outer edges of the hive. Progressively replace each one with clean drawn comb or foundation, but at all times, take care not to split the brood nest. If combs that are to be replaced contain stores they can be utilised by the colony if they are laid horizontally over the feedholes on the crown board. The bees can be encouraged to take down the food if areas of the cappings are scraped away to expose the honey.

- **Re-assemble the hive**

After completing your inspection reassemble the hive as quickly and gently as possible ensuring that the edges of the boxes are squared up and there are no holes or spaces through which robber bees and wasps can enter. Replace the roof in its original orientation and complete your records.

FEEDING BEES

The replacement of harvested honey, removed by the beekeeper, requires the consumption of 25% of the sugar fed to it just to provide the energy required to process the syrup into a form in which it can be stored.

The nutritional status of colonies and the feeding of bees is an important part of honey bee health management. Typical situations where feeding may be necessary are as follows:

- Rescuing a starving colony, for example in a poor year for forage.
- Encouraging the growth of the colony in spring.
- Stimulating the queen to lay.
- Feeding nuclei or swarms.
- In preparation for winter.

- Supplementing overwinter stores.
- Where a syrup solution is used as a treatment carrier for a drug, e.g. Fumidil®B.

Nectar substitutes

Sugar syrup should be made from pure white and refined granulated sucrose sugar. Brown sugar should not be used as it may cause dysentery.

Syrup for stimulating the queen to lay, the workers to feed royal jelly, for feeding nuclei or swarms, or the administration of drugs should be made by dissolving one part sugar into one part water by weight. For maintenance feed during autumn and over winter two parts sugar to one part water by weight should be used. The sugar will dissolve quickest in hot water but it must be only lukewarm when fed to bees. Some beekeepers add one level teaspoon of salt per 4.54 litres (1 gallon) of winter feed syrup, or sprinkle the salt over the frame tops in an attempt to control Chalkbrood.

High fructose corn syrup (a converted starch product) can also be used for feeding in the autumn but it is not universally accepted as good practice. Other proprietary liquid feeds specially prepared for feeding bees are available from beekeeping equipment suppliers; follow the instructions given with the product.

Fondant is an excellent food supplement in a long winter and is
placed on the frames above the cluster.

Sugars can also be fed as a fondant paste placed across one of the holes
in the crown board to allow bees to access it. In emergency situations it
should be placed on the frames, directly over the cluster. It is possible to
make fondant yourself, but it is a messy process. It is better to purchase
a proprietary product such as Apifonda®, which is ready to use fondant
paste consisting of refined sugar (sucrose), fructose and glucose. It is
packaged in a plastic bag, cut it open when in the apiary and slice off
the required amount of fondant.

An emergency method is to use a bag of granulated sugar, puncture
it in several places on one side and on the same side cut a 2cm diameter
circle out of the bag to expose the sugar, add enough water through the
hole just to moisten the sugar and place the bag with the exposed sugar
face down over the feed hole of the crown board so the bees can gain
access to it.

Types of feeders

Feeders come in three basic types, contact, fast/rapid and internal
feeders. Typically a contact feeder is an upturned container filled with
syrup, with a lid in which a perforated mesh is inserted. The feeder
is surrounded by an empty brood box to support the hive roof. After
filling and sealing, the feeder must be inverted over a container to
collect any spilt syrup and draw a vacuum, if there is a break in the

vacuum spilt syrup will encourage robbing. The bees can access the feed through the mesh, and thus the syrup is immediately available to them. Contact feeders should be replenished away from the hive. Bees will often propolise the gauze of these feeders preventing access to the syrup by the bees. If this happens remove the feeder and use boiling water or a boiling solution of washing soda to clean it. An alternative is to immerse the gauze in industrial alcohol.

Fast / rapid feeders are placed over the holes in the cover board and bees gain access to the syrup, and are prevented from drowning in it by a perspex cover. Feeder types such as Ashworth, Miller and Brother Adam cover the whole cross section of the hive and do not need a box to support them. Smaller types of fast / rapid feeders require boxes as above. Bees are sometimes reluctant to work these feeders in cold weather because they do not have direct contact with the syrup.

Internal feeders are frame sized and have a reservoir which can be filled through the top of the frame. The bees gain access to the feed through an aperture along the top side of the frames. An alternative that does not require a feeder is to use uncapped frames of liquid honey inserted into brood boxes. The uncapping makes the honey more accessible to the bees. Ensure that the honey comes from a disease free colony.

When feeding:

- Ensure that there is no spillage or dripping of syrup in the apiary.
- Prevent robbing through ensuring the hives are bee tight and reduce hive entrances.
- Ideally feed in the evening just before dusk.
- Take steps to prevent sugar syrup finding its way into supers and then into honey which will be sold. One way is to clear and remove the supers before feeding.

As a general rule try to get the feeding completed by the end of September, in a poor forage year this will not be possible and you may have to feed with fondant well into February or March.

Protein feeding and pollen supplements

Protein is crucial to the development of bees; if pollen is in short supply adult bees will be short lived. Feeding protein in early spring can be used to increase the hive population, improve their health status and strengthen the colony in preparation for queen rearing. Protein feeding in the late summer / early autumn can be used to increase the protein status of the over-wintering bees, which is crucial to the survival of their colony. If the cluster is positioned over comb without pollen stores or stores are running low, the bees can mobilise the protein stored in their fat bodies to feed the brood produced early in the following year. This is especially important during colder and wetter springs when pollen availability may be reduced.

Protein can be fed in the form of pollen supplements. With pollen collected from pollen traps and refrigerated until required then mixed with a range of substances.

An example of a suitable pollen supplement contains:

Pollen	150g
Lactalbumin or caseinates	328g
Brewers' yeast	656g
Sugar	186g
Water	450ml

Mix the pollen and water, add the yeast and then slowly add the lactalbumin. As the mixture thickens, add more water until a firm consistency is achieved. Form the mixture into patties and store it between two sheets of wax paper or cling film in a freezer at -18°C to kill wax moth eggs and larvae (possibly present in home harvested pollen).

Ideally when using the patties place them close to the unsealed brood, on the frame tops directly over the brood, or smeared into cells of an empty drawn frame. After one week check and remove any remaining patties that have developed mould and fungal growth, and replace with fresh patties. Ensure the colony has access to plenty of sugar syrup or honey, as the workers bees will be expending more energy in caring for the increased brood produced as a result of feeding with protein supplement.

If, on inspection the pollen patty is not being consumed, the colony may be queen-less, or the bees are obtaining sufficient natural pollen for their needs.

Pollen substitutes

These differ from pollen supplements in that they do not contain any pollen. Unless properly formulated they may lack essential vitamins, amino acids and minerals. They are more readily accepted when natural pollen is added. My advice is to buy ready prepared products from a supplier.

Water

Bees will collect water locally from ponds, ditches, drains and taps on water butts. If none of these are available nearby an alternative is a shallow container, such as a plant pot saucer, containing gravel. The saucer is filled with water to just above the gravel to avoid drowning the bees, and placed in the apiary in a warm sheltered spot. In hot or dry weather it should be checked regularly and not allowed to dry out.

SECTION 6

CARING FOR THE QUEEN

The queen is the key individual in a honey bee colony, it is vital that she is nurtured and maintained in peak condition.

FINDING QUEENS

Beekeeping often requires you to find the queen and keep her in a safe place during a manipulation. Care and a steady hand are necessary; have a queen cage and other equipment ready before you start. Place her into the cage over the open hive so that if you drop her she will fall into the frames. Marked queens are easier to find as long as the marker colour has not worn off, although experience shows that when reading and examining the combs you know where to look for her. Use dummy boards to create space and also remove the next frame to enable you to move the combs along the runners and have a good chance of seeing the queen before she runs away from the light. Using smoke often provokes this response so use a fine water spray as a substitute to help find her.

During an inspection a queen can drop off the frames and seek refuge in the hive corners, this makes her very difficult to find. If the queen falls to the ground wait a few minutes and stand still so you do not tread on her; she will attract worker bees, so look for a small cluster within which you will find her.

If she is not obvious on the frames they can be paired up leaving space between each pair. After a few minutes when the bees have returned to the frames, open the pairs of combs and look for the queen on the inside faces.

If it is essential to find the queen, and all other techniques have failed, a queen excluder can be used to sieve her out and have a queen holding cage ready. The bees are shaken through the excluder, held down by an empty brood box or super, into a brood box containing a few frames. The workers pass through the queen excluder and the drones and queen are retained above it. This technique may not work at swarming time if the queen has already been slimmed down and can pass through the queen excluder.

QUEEN NUTRITION

Queen nutrition is complex and includes a diet of brood food and royal jelly which varies throughout her development. Royal jelly is a thick, creamy, highly concentrated source of proteins, essential amino acids, fats, vitamins and other nutrients with a moisture content of 65–67%. Its nature and composition changes during the life of the queen and during the queen's development it includes food from both the mandibular and hypopharyngeal glands. The molecular mechanisms underlying the action of royal jelly are unknown.

During the first three days of the life of a larva selected to become a queen the food given is mostly 'white food' from the mandibular glands of the worker and nurse bees; it is 34% sugar consisting mainly of glucose. On the third day the sugar concentrations become even higher. On days 4 and 5, the last days as a larva, the food given is in a ratio of 1:1 of white from the mandibular glands and clear from the hypopharyngeal glands, both produced by the workers. The actual protein and amino acid requirements are unknown, even though we have some idea of the composition of royal jelly. Some royal jelly proteins exert context-dependent functions. Those used in the brain development and function have different effects to those used in the queen's reproductive development.

Protein and amino acid requirements of adult queens after emergence are also unknown. When the queen bee first emerges from her cell she eats honey and pollen directly from the comb and after a few days the workers start to feed her. As the queens develop they are fed a diet containing protein consisting of royal jelly and honey produced by older worker bees and this feeding continues throughout the year. A high protein diet fed to the queen is essential because she requires protein for the development and maturation of her reproductive organs. The production of eggs and pheromones also requires a high quality diet. The queen continues to grow and mature after her emergence.

QUEEN REPLACEMENT

The increasing incidence of poor mating results in the queen receiving insufficient numbers of healthy sperm. This restricts her ability to lay enough fertilised eggs to maintain large colonies for more than two years. Good beekeeping requires colonies to have healthy young queens able to produce large quantities of fertilized eggs. To achieve this, a programme of regular queen replacement every two years should be incorporated into your bee health management strategy.

The discovery of a drone breeding queen or the accidental killing of a queen during a colony inspection can happen throughout the season. It is possible to re-queen at any time during the active season when queens are available, usually between March and October; the best time is generally mid August to early September in warm weather with a good nectar flow or when you are feeding the bees.

Re-queening can occur naturally as follows:

- Through swarming, when a new queen is produced in the original colony and the old queen leaves with the swarm.
- By supersedure.
- As a result of the production of emergency queen cells, usually when the queen is lost or damaged.

The queen is the mother of the colony, responsible for many characteristics exhibited by its bees, some good and some which should be eliminated. For the benefit of future beekeepers we should always seek to improve the quality of our bees and, like other aspects of beekeeping, queen rearing requires some forward planning.

The following colony traits can be selected for improvement through queen rearing:

- Good temperament.
- Productivity.
- Reduction in the 'following tendency'.
- Disease resistance.
- Swarming tendencies.
- Longevity.
- Comb building activities.
- Hygienic behaviour.

These are determined by the genetic make-up of the parents, but it is easier to select queens with desirable characteristics compared to drone selection. Manipulating the genetic impact of the drone is much more difficult and can be done only through artificial insemination of queens, or the use of isolated mating stations where the drones present are selected and bred in large numbers.

Re-queening can be carried out in a number of ways.

- As a result of a controlled programme of queen rearing, mating and introduction by the beekeeper.
- Replacing the queen with one obtained from a beekeeper local to your area.
- Ready mated and tested queens bought in from specialist suppliers.

Rearing new queens

If you only have a small number of colonies to be re-queened I would recommend using swarm cells and the artificial swarm technique. This retains the old queen and enables foragers to continue producing honey whilst the queen cells in the old colony are used to produce new queens.

If you require a larger number of new queens from one or two selected breeder queens, and there are inadequate numbers of swarm cells, you must use a technique to produce larger numbers of larvae raised in a separate cell-building colony. Sealed cells or immature virgins can then be moved to mating nuclei.

Queen-rearing involves:

1. Selecting the genetic material to be propagated (eggs, larvae or queen cells containing larvae from the selected breeder queen).
2. Raising the eggs or larvae in queen cell building colonies. These are used to feed and care for a caste-undetermined female larva to ensure she develops and emerges as an adult virgin queen when placed as a sealed queen cell in a mating nucleus colony.
3. Mating the virgin queens.
4. Evaluating the new mated queens.
5. Introducing them into colonies.

Producing good queen cells from larvae

Having identified the queen whose characteristics you wish to select, it is essential that new queen cells are raised in the best possible conditions. Plentiful food is the critical element in determining whether the larva will become a queen and develop to her full potential. A large number of young nurse bees must be available to ensure that all the larvae to become queens are fed abundantly with the correct food from the time the eggs hatch until the mass-provisioned larvae pupate.

The selection of larvae of the correct age is the key to successful queen rearing using larval grafting. The selected larvae should be:

- Almost transparent in colour.
- Less than 2mm long.
- Nearly straight (only slightly 'C' shaped).
- Ideally about 12 hours old.
- Definitely less than 36 hours old.

A colony normally produces queen cells at the peak of its strength under optimum conditions.

These include:

- Abundance of high quality food, i.e. pollen, nectar and honey.
- Abundance of nurse bees.
- An excess of royal jelly.
- A large number of wax secreting workers building comb to store pollen and nectar brought in by a large foraging force.

These conditions can lead to swarming unless the beekeeper exercises swarm prevention and control techniques.

Queen rearing from swarm cells

When only a few queen cells are required from colonies with desirable characteristics, naturally built swarm cells resulting from preparations for swarming can be used. The optimal conditions described above ensure that the queen larvae will be successfully reared by the nurse bees and well provisioned before the cells are capped. The sealed queen cells are then cut from the combs and placed in de-queened colonies, or nuclei. From these the virgin queens will emerge, fly and mate.

Queen rearing outside the swarming season

To raise good queen cells, other than during the swarming season, it is necessary to simulate the natural factors that occur at swarming time. This may include adding young bees and emerging brood taken from other colonies. When there is an abundance of pollen in the combs, or a good nectar flow the production of brood food can be increased by feeding light sugar syrup. In the absence of adequate pollen and nectar flows, pollen supplements and sugar syrup can be fed.

New queens resulting from supersedure

Queen cells produced under supersedure, usually when the queen is failing are probably less reliable because they have been mated under sub-optimal conditions. Supersedure can occur at anytime during the early spring, mid summer or autumn.

New queens resulting from emergency queen cells

Colonies will also produce queen cells under emergency or queen-less conditions. This behaviour is used in queen rearing by removing the queen from the colony and allowing the colony to re-queen itself by producing a number of virgin queens. These can be harvested by the beekeeper and raised in other colonies such as nuclei. They are usually of inferior quality to prime swarm cells.

Some other techniques

When the breeding queens and the queen cell-building colony occur in separate hives, larvae (or eggs) are obtained from the colonies of the selected breeder queens and raised in the queen cell-building colonies. In one technique the larvae are obtained by inserting into the colony of the breeder queen, frames holding specially cut pieces of foundation that the bees draw out and in which the queen lays. The comb is removed and the egg-containing margins of the comb are trimmed away so that the youngest larvae are at the new comb edge. The comb is then ready to be put into a de-queened, queen cell-building colony.

In a second technique, larvae, preferably less than 24 hours old, but no more than 3-4 days after the egg was laid, are transferred from the brood frames of a selected queen breeding colony and grafted into a frame capable of holding a number of queen cups. This frame is then placed into the queen cell-building colony where the workers feed and rear them. From here the virgin queens in their sealed cells can be distributed into mating nuclei.

The Jenter method is a modification of this technique where the queen is confined for a period of 24 hours on an area of artificial comb containing small, early stage, plastic queen cell plugs into which she will lay her eggs. After 24 hours the queen is released and the individual plugs containing her eggs are removed and inserted into the other parts of the artificial queen cell to form queen cups, these are then placed into a frame that will hold a number of them, preferably less than ten. The queen cells are then raised and distributed as previously described.

Take care when handling frames

Frames containing developing queen cells must not be shaken during manipulation, especially when sealed cells are present. Queen cells hang down from the face of the comb and the larvae are embedded in the food at the top of the cell. Any downward shaking will dislodge the larvae from their food, they will lose contact with it and die, and valuable time will be lost waiting in vain for them to emerge.

Aspects relating to drones

The drone culling technique will, if carried out as recommended for Varroa monitoring, seriously deplete the number of drones in a colony and reduce the number available to mate with virgin queens. It will also potentially deplete the genetic diversity achieved by matings in the drone congregation. It may be beneficial to have a colony in which drone foundation is added to boost the drone population, although it could increase the population of Varroa mites.

Drone mother colonies i.e. colonies selected to boost drone numbers should be treated for Varroa 40 days prior to when the drones are required for mating. Formic acid should not be used for Varroa control at this time as it stimulates the workers to remove the drone eggs from the colony.

BUYING IN QUEENS

Licences are required for the importation of queens and attendant workers into the British Isles. Further information on the restrictions and procedures to be followed when importing queens into the British Isles can be obtained from the UK National Bee Unit. Often it is simpler to source queens from a number of origins and selective breeding programmes in the British Isles.

Buying in queens is quite an investment. The advertised costs for mated and tested queens vary depending on the type of queen being offered. There are often small additional charges for marking the queen and clipping her wings.

Available races and strains of queens

The honey bees of the British Isles are a mixture of genetic types produced through the introduction of strains of bees from abroad, the Dark European Honey bee and its variants, and those produced through crossing with imported strains and races of honey bees. There are suppliers of queens who offer specific strains and races, usually maintaining their integrity through techniques such as artificial insemination and mating management.

Types of queens available in the British Isles include Italian, Carniolan, Caucasian, Dark European Bee, and strains adapted for local situations bred from mixtures of one or more of the types mentioned above.

Making general statements in beekeeping is always risky however the following comments have been made about some of these types.

Italian bees *(Apis mellifera ligustica)* have the following characteristics:
- Colonies usually large and overwinter well.
- Very good honey producers.
- Usually gentle and non aggressive.
- Swarming instinct not especially strong.
- Use minimal propolis.
- Keep a clean hive and are quick to get rid of wax moth.
- Queens lay throughout summer so a large proportion of their stores are used for brood rearing.
- Have a strong robbing tendency.
- Yellow coloured bands on the abdomen.

- Bees sit quietly on the comb when frame is manipulated.

Caucasian bees *(Apis mellifera caucasica)*
- Very gentle bees.
- Do not swarm excessively.
- Brood build–up is in late spring.
- A good but not exceptional honey producer.
- Collects and uses a great deal of propolis.
- Brown in colour.

Carniolan bees *(Apis mellifera carnica)*
- Very gentle bees.
- Probably the best over- wintering bees.
- Little use of propolis.
- Rapid build–up in the spring.
- Summer brood rearing responds quickly to pollen and nectar flow.
- Usually not inclined to rob.
- Tend to swarm more, possibly due to rapid spring build–up.
- Not as productive as Italian strain.
- Greyish in colour due to hairiness.
- Bees sit quietly on the comb when the frame is manipulated.

Dark European Bee *(Apis mellifera mellifera)*
- Slow development in the spring.
- Moderate brood production throughout the season.
- Low consumption of food stores.
- Good longevity of workers.
- Compact pattern of brood and stores.
- Extensive use of propolis.
- There is strain variation in which the bees keep the hive floor clear of wax and debris, limiting the amount of wax moth found in the colonies.
- Nervous on the comb and will cluster on the edges of frames during manipulations.

Your choice of type and timing of queen replacement is influenced by a number of factors:

- Your local climate.
- The forage crops available locally, especially the timing of peak nectar flows and periods of no nectar.
- Your beekeeping strategy e.g. you want to maximise honey production.
- The apiary location e.g. if you keep bees in a suburban garden you may wish to use gentle bees.

Some suppliers offer locally adapted bees, usually those they use in their own commercial operations. They may claim their queens are derived from or similar to the Dark European Bee.

Purchasing queens from reputable local sources is probably the best route for beekeepers with limited experience of queen rearing. Experimentation with other types can be made when the beekeeper has more experience and is able to cope with aggressive colonies of bees which can result from some crossings in subsequent years.

Advantages of buying in queens

Queens of known type can be purchased and will have been mated and tested. These queens will raise colonies with predictable behaviour.

Disadvantages of buying in queens

You need to re-queen each year or every other year with queens from the same source or type otherwise crosses between next generation queens and the drones of the local type may produce very bad tempered bees.

If inadequate precautions and techniques are used in the queen's introduction to the colony she may be lost. Buying in a replacement will be expensive.

QUEEN MATING

Successful mating is very dependent on good flying weather. The temperature at the time of the mating flight must be at least 16°C, and preferably above 20°C, with little cloud cover and wind speed less than 20-28 km/hr. Queens mated in less than ideal conditions are often superseded, probably because of the inadequate transfer of semen during mating flights. Most queens make one or two short orientation flights followed by between one and five mating flights over a period of two to four days. Current thinking is that the queen continues mating flights until her spermetheca is full of semen.

Mating flights take longer in early spring than in summer. If bad weather prevents the queen from taking mating flights for three to four weeks after her emergence, she will degenerate and begin to lay drone eggs. Remember to record the likely date of emergence and note the subsequent weather for the next few days. Poor weather is likely to lead to poor mating and the prospects of drone laying queens are very real if the queen is inadequately mated.

Ripe ready to emerge queen cells or virgin queens can be transferred to different types of queen-less colony in which they can emerge, fly to mate and begin their egg laying.

These include:

- Micro-nuclei e.g. polystyrene Apidea® and Warnholz mininucs.
- 3 or 5 BS frame nuclei.
- A brood box divided into 3 or 4 compartments each with a separate entrance.

An advantage of using full sized frames from a disease free colony is that they can be selected with good honey and pollen stores which will help the new colony in the event of poor foraging conditions and also nourish the new queen in her development. Another significant advantage in mating queens in colonies using standard BS brood sized frames is that it is possible to better assess the qualities of the queen before she is introduced into the colony to be re-queened. An evaluation is more difficult using micro-nuclei such as the Apidea® or Warnholz mating box which can only support small colonies. When using micro-nuclei boxes the mated queen must either be introduced into the colony to be re-queened, or placed in a prepared nucleus colony once she starts egg-laying. Mating nuclei can be established in the same apiary in which the queen cells were raised.

Queens usually return fertilised and healthy from their mating flights carried out from their apiaries or close to where they were raised. By contrast at least a third of queens which are moved to another location some way away and mated or which are mated from mini-nuclei are lost. The reason for this may be the inadequate size and abilities of the groups of workers which fly out with the queen and appear to accompany and guide her on her mating flights. Added to this is the fact that the queens either have no knowledge or only a limited knowledge of the area outside the hive gained during the short period of their orientation flights prior to starting their mating flights.

INTRODUCING THE QUEEN

Having made the investment of purchasing a queen or spending time and effort rearing your own queens, you need to take a number of precautions before introducing the mated, and if bought-in probably tested, queen into a colony. The state of the colony at the time of the queen introduction is of vital importance.

- Never attempt to re-queen a colony involved in swarming preparations or complications arising from swarming.
- Never introduce a queen into a hungry colony. If there is no nectar flow in progress feed the colony generously with sugar syrup 2-3 days before and 2-3 days after introducing the new queen.
- Check that there is no virgin queen already present in the colony.
- If there is a drone-breeding queen present, find and kill her before the new queen is introduced.
- Never remove an actively laying queen and try to replace her immediately with a travelled queen that may not have laid eggs for a number of days.
- Never introduce a queen to a colony consisting entirely of old bees that has long been queen-less. There is a need to add brood and young bees before introducing the new queen.
- Never introduce a queen to colonies with laying workers as they will be very hostile to new queens.
- A queen is usually accepted by a colony that contains mostly sealed or emerging brood.

A good nectar flow helps with the introduction of a new queen, but if there is no flow the colony to be re-queened should be fed a light strength syrup (mimicking nectar) until she is laying well.

Cage methods of introduction

As a rule queens can be introduced most successfully if the colony to be re-queened is made queen-less, and the new queen is introduced by a cage method at the same time. The bees in the de-queened colony will miss their queen within a few minutes and you will become aware of their characteristic behaviour and agitated buzzing. Ideally valuable queens should be introduced to young bees in a nucleus colony before the nucleus is united with the colony to be re-queened.

A bought-in queen will usually be supplied in a postal queen cage accompanied by six workers. These workers should be removed and replaced by a similar number of workers from the new colony or nucleus. It is important that these workers are well fed with syrup, honey or nectar. The queen and attendant workers then can be transferred to a queen cage. Close the ends with a small single sheet of newspaper, secured with elastic bands and then introduce the cage into the queen-less hive. Suspend the cage by pressing it gently into the face of a brood comb. After two weeks open the hive, check that the cage is empty; the workers have eaten through the newspaper to release her, and examine the comb for the presence of eggs and larvae.

Typical queen introduction cage inserted between brood combs.

Direct methods

These methods are not without risk.

1. A colony that has been de-queened and the bees then shaken onto a sheet on the ground in front of the hive will often accept a new laying queen that is dropped into them as they move back into the hive.

2. After a colony has been made queen-less the new queen, preferably taken from a nucleus hive a few minutes before, is run into the entrance of the hive followed by some good puffs of a cool smoke. The colony should be fed before and after the introduction of the new queen.

3. A third method involves the use of a dilute sugar syrup spray. After de-queening immediately mist-spray the sides and tops of the frames in the brood chamber and the queen in her introduction cage. Release the queen directly onto the top of the frames and spray again as she descends into the brood box, immediately close the hive. This method should not be used in circumstances where robbing is likely.

Direct transfer of combs and bees from a nucleus colony

If you have reared and mated the queen in a nucleus hive using full sized frames, the colony can be united with a de-queened colony. Apply a spray of light sugar syrup over the sides and tops of the frames to be introduced from the nucleus to divert the attention of the bees in the receiving colony. A number of frames, preferably brood-less, are removed from the de-queened colony and replaced with the same number of frames of brood, with the new queen on the inside face of one of the combs. The frames from the nucleus should be placed in a block at the outer edge of the brood chamber.

An alternative method is to de-queen the colony to be re-queened, transfer the nucleus into a full sized brood box and place it on a sheet of newspaper, punctured with small holes, on top of the de-queened colony. The bees chew through the paper, and intermingle. Select the best combs and place them in the lower brood box and remove the top box. It is advisable to check that there are no queen cells in the brood of the former queen, if so remove them. Unless there is a very strong nectar flow colonies being re-queened should be fed until the new queen is well established.

QUEEN MARKING AND WING CLIPPING

Traditionally queens have been marked with a colour that changes every year over five years. If new queens are marked with the colour of the year in which they were hatched you know their age. Today's beekeeping tends to require a more rapid turnover of queens, so in my opinion three colours are enough to enable the beekeeper to keep track of the age of his queens.

The queen needs to be carefully handled. Queen catching and holding devices can be purchased from beekeeping suppliers. Beekeepers with a steady hand and good eye sight should hold the queen by her thorax using the thumb and forefinger of the left hand (if you are right handed) and apply the colour to the top surface of the thorax. Use marking pens or paint sold specifically for marking queens otherwise you may damage the queen or provoke the workers to reject her. Allow the colour to dry before returning the queen to the colony. Ensure that no colour is applied to the antennae, the sides of the thorax or the base of the wings or legs.

Surrounded by her attendants this laying queen is
marked on her thorax.

Do not be tempted to mark or clip the wings of the queen before
introduction, or even a few days after, as this can trigger rejection or
induce supersedure. You can wait until the following spring to mark
the queen and clip her wings if you wish. Never clip the wings of a
virgin queen before she is mated, as she will be unable to fly and mate.
The fore and hind wing of one side are clipped (or cut) with a sharp pair
of scissors to about halfway along their length. Do not cut any further
or you will damage the costal vein and haemolymph will be lost. If a
leg is accidently cut as long as it is not one of her forelegs which she uses
to measure the cells and determine whether the egg to be laid will be
fertilised or not, the colony may continue to accept her if there is no
other damage.

TABLE 19 Queen marking colour codes

For years ending with	Colour
1 or 6	White
2 or 7	Yellow
3 or 8	Red
4 or 9	Green
5 or 0	Blue

QUEEN TESTING AND ASSESSMENT

Having selected your breeder queen, reared the new queen and introduced her, unless you are simply replacing an old queen you will want to know whether you have succeeded in improving quality and eliminating the undesirable characteristics of the former queen.

Often a spell of poor weather or the cessation of the nectar flow will coincide with the introduction. She may not start egg-laying or the workers may reject her if you open the colony too soon or during a period of forage shortage. Always give the new queen plenty of time to settle down and start egg-laying and observe the behaviour of her progeny before beginning the assessment. It will take some time for them to become numerous enough for the colony to take on her characteristics. Realistically it is either later in the season or early the following season before you can make a thorough evaluation. This is where the benefits of a good colony record system are invaluable.

A queen whose brood shows one or more of following characteristics should be replaced:

- A scattered or spotted brood pattern with many cells within the brood area that do not contain an egg is an indication of reduced egg viability or inbreeding. This occurs when the queen has mated with drones that are closely related to her. This results in diploid drones that may comprise up to 50% of the brood. Normally the workers eat the diploid eggs and larvae.

- The queen continues to place several eggs in a single cell after a period of 7-10 days. New queens may initially lay multiple eggs in a single cell, sometimes on the cell wall but after a few days they will lay normally.

- Drone brood in worker cells, or multiple eggs in a cell may indicate a drone layer.

- Half-moon syndrome is a condition where some of the symptoms are similar to EFB. Many cells will contain multiple eggs attached in chains and laid on the cell wall and often drone eggs are laid in worker cells. It is thought that the condition is due to poor nutrition of the queen after emergence.

- Non hatching eggs which may be sterile. Sterile eggs are occasionally found where a newly mated queen begins and continues an apparently normal laying behaviour, but the eggs do not hatch. This condition is probably genetically based and due to inbreeding.

- Reduced egg laying that may indicate an abnormality.

THE MANAGEMENT OF SWARMING

SWARM PREVENTION

Swarming is the natural process by which colonies of bees reproduce. Much time and effort is spent by beekeepers in trying to prevent bees swarming, but a better approach is to assume all your colonies are going to swarm and plan how to deal with them. The skill is to know when this will happen and then to manage the situation in a way that is acceptable to you, does not frustrate the colony, and enables the bees to fulfil their natural instincts.

Today beekeepers use techniques to try to prevent their colonies from starting swarming preparations, including being aware of overcrowding and adding supers well before this happens. Add the first super as drawn comb and then add the next one as foundation at a later date thereby ensuring the bees have sufficient space in which to place and process nectar. This may delay swarming and give you time to plan your swarm control management. For this it is important to understand the principal internal factors causing swarming, these include:

- Brood nest congestion resulting in restricted space for the queen to lay eggs.
- Reduced transmission of queen substance because of colony overcrowding and restriction of the queen's movement through the colony.
- A high proportion of young workers.
- Older queens producing lower quantities of pheromones.
- Some colonies have a greater tendency to swarm as a result of a genetic trait.

There are also external factors such as the weather which cannot be controlled by the beekeeper. The swarm stimulus in spring is promoted by a good nectar flow irrespective of the quality of the pollen.

Colony inspection techniques for assessing swarming status

Early to mid April is a good time to start your inspection regime for monitoring colony preparations for swarming, although this depends on your local climate and the colony conditions prevailing at the time. Before carrying out your apiary inspection during the swarming season prepare to deal with one or more colonies that may be about to swarm in case you have to deal with them immediately.

When checking for swarming preparations look for the following things:

- Drone brood on the combs and a good number of mature drones flying; monitor the increase in drone brood.
- Does the queen look thinner than in previous inspections? In other words have the workers slimmed her ready for swarming?
- The absence of eggs in worker and drone cells with the queen still present. She will cease to lay prior to swarming.
- The start of queen cell preparation. In single brood boxes they will usually be found along the top, bottom and side edges of the comb; in double brood systems they hang from the lower edges of the combs in the upper box.
- The presence of eggs in queen cells.
- Sealed queen cells, these can trigger a swarm 8 - 9 days after the egg was laid, if the weather is suitable.

Check that your queen is marked and, if you wish, clip one of her wings. In colonies where the queen is unclipped the swarm may go earlier if the colony is disturbed and the weather is suitable even if the queen cells, containing the developing virgin queens, are unsealed. In practice this means that a colony headed by an unclipped queen requires inspection every 7 days. This can be quite an onerous task but is necessary if swarming is to be prevented. Clipping queens does nothing to prevent colonies swarming, but it has the advantage of enabling the beekeeper to extend the period between inspections to 10 days.

Clipping the queen's wings prevents her flying with the swarm. She may emerge from the hive as the swarm issues, but usually will fall to the ground in front of the hive; she may crawl back into the hive or reach the underside of the floor. The swarm will either return to the parent colony, or some of it may cluster around the queen on the ground. She and the cluster can be recovered from such situations either by placing a skep over the cluster, or a nucleus box containing a frame of comb beside it to attract the bees into it.

The 10 day inspection

There are several key dates in a queen's development after the egg is laid:

Day 3   the eggs develop into small larvae

Days 8/9   the larvae are sealed in their cells

Day 16   the virgin queens emerge

The 10 day inspection is based upon the following rational. If, when you carry out your inspection, the colony is not producing queen cells, the earliest date a swarm could issue from a colony with a clipped queen is 17 days after the egg was laid. If the queen is clipped she is not likely to leave the colony until the first virgin queen is ready to emerge. In practice a colony with a clipped queen can be inspected on a 14 day basis until queen cells containing eggs are observed.

If when carrying out the inspection the colony is seen to be raising queen cells and these are destroyed by the beekeeper, the oldest worker larvae that can be selected to replace those killed are 7 days old (i.e. 3 days as eggs and 4 days as larvae). These should emerge as virgin queens 9 days later and can be flying as part of a swarm by the tenth day after inspection. However bees usually select a worker larva at least 5 days old (3 as an egg and 2 as a larva) so the virgin queen will not emerge for at least 11 days.

In the meantime the clipped queen may emerge and may be lost but the swarm will go back to the hive from which it came. However if the swarm emerges on the sixteenth day after inspection it will contain at least one virgin queen. Unless its emergence was seen by the beekeeper and it was successfully caught and re-hived the swarm will be lost.

SWARM MANAGEMENT OPTIONS

Having identified colonies that are about to swarm you have two options:

1. Swarm management with no increase in colony numbers. This is usually used to re-queen a colony without loss of brood continuity.
2. Swarm management with an increase in colony numbers.

This can be used to provide several new queens to re-queen other colonies or produce new colonies.

In both options there is a need to prepare an artificial swarm. The artificial swarm is the colony that contains the original queen; whilst the parent colony is the one in which the new queen cells are being raised and from which the virgin queens emerge.

Have all the equipment you require ready in the apiary. Carefully read through the description of the technique you intend to follow and make notes on an index card. Have a practice run so you know where to put things and in which sequence.

Preparing the artificial swarm

1. Gently smoke the entrance of the hive and stand aside for two minutes.
2. Remove the roof, gently smoke over the top of the crown board, and place the inverted roof to one side of the hive.
3. If the colony has one or more supers above a queen excluder, insert your hive tool between the top of the queen excluder and the bottom of the super.
4. Remove the super(s) and set to one side on top of a second crown board. Cover the top of the supers with a cover cloth; this will help to keep the bees in the supers.
5. Place the original brood box and floor on a hive stand about 1 metre away from its original site.
6. On the original location of the colony place a floor with an entrance and an empty brood box.
7. Gently smoke over the frames in the brood box containing the queen (original colony). Use the minimum amount of smoke to avoid panicking the queen off the face of the combs and into the recesses of the hive.
8. Carefully go through the combs until you find the one on which the queen is located.
9. Use a queen marking cage to hold her on the comb, while you check the rest of the comb for queen cells.
10. Relocate her to another frame if the frame she is on is the only

one with queen cells. There must be no queen cells on the frame containing the queen.

11. Place the comb and its bees with the caged queen into the brood box on the original site and release her.

12. Make up the rest of the brood box with drawn comb or foundation.

13. Replace the queen excluder, supers (if relevant) and the crown board over the brood box and replace the roof.

Many of the flying bees will relocate from the parent colony back to the original site where the brood box containing the artificial swarm is now positioned. Unless there is a good nectar flow it is necessary to feed the artificial swarm with sugar syrup.

With preparation and organisation the following options are available:

Swarm management with no increase in colony numbers

Here the objective is to maintain the honey production of the colony through keeping the foraging bees as a single colony rather than losing them in swarms and subsequent casts. In this technique the rearing of brood continues uninterrupted, the artificial swarm will continue to forage for nectar and the colony has been re-queened.

1. Move the parent colony close to one side of the artificial swarm.

2. Select two unsealed queen cells containing larvae and destroy the rest.

3. Mark with a drawing pin the top of the frame that has the selected queen cells.

4. Calculate the number of days until you expect the virgin queens to emerge and the number of days by which the virgin queens should have been mated and begun egg laying. This can be up to 21 days after the date of emergence; if there is no sign of eggs after 21 days the virgin queen has either been lost on her mating flights, or is damaged in some way. The bad-tempered behaviour of the bees will tell that there is no queen present, even though you cannot see her.

5. When the new queen is laying eggs you can unite this colony

with the artificial swarm, having first found and dealt with the original queen by killing her unless she is valuable and you wish to use her for breeding purposes. In this case transfer her to a freshly made-up nucleus colony.

6. The supers are placed on top of a queen excluder on the top of the uppermost brood box. See Section 9-Uniting colonies.

Another method that also results in no increase in colony numbers is as follows:

Instead of placing the original colony to one side of the original site it can be placed over a Snelgrove Board on top of the roof of the colony containing the artificial swarm and its supers. The board has its own entrances. Open one of the entrances, preferably the one on the opposite side to that of the artificial swarm. Then follow the same queen cell selection, rearing and mating techniques as described in Section 6.

When the new queen is mated and laying and you are satisfied with her progeny, you can remove the Snelgrove Board, and set aside the supers, find and deal with the original queen in the lower box as above. Then use the newspaper uniting method to combine the two colonies. The supers are then replaced over a queen excluder on top of the uppermost brood box. The original queen can be retained in a newly made up 2-3 frame nucleus colony if you wish to keep her.

Swarm management with an increase in colony numbers

This can be achieved by producing two nuclei to grow into new colonies.

1. Place two nucleus boxes on their floors on the roof of the artificial swarm. They should be positioned side by side with their entrances facing in opposite directions.
2. Go through the combs of the original colony and split it into two equal parts each having two unsealed queen cells.
3. Remove all the other queen cells.
4. Mark the frames containing the selected queen cells.
5. Replace the crown boards and the roofs of the nucleus boxes.
6. Calculate the time after which the virgin queens will have emerged, mated and should be laying eggs.
7. Once the queens are laying the colonies can be transferred to

full brood boxes and moved to another site to continue their development.

Swarm management to produce queens for re-queening colonies

This is an excellent method to produce several queens from a queen heading a colony that has good characteristics, e.g. temperament.

1　Adapt a brood box and floor to make several compartments (up to 4), each with their own entrance, internal partitions to separate the compartments, and individual crown boards.

2　The adapted brood box is placed on top of the roof of the artificial swarm with its own floor.

3　The frames of the original colony are removed from their brood box and a frame containing up to two unsealed queen cells is placed into each compartment.

4　The other frames are inspected, their queen cells removed, and the frames and bees distributed, ensuring there is food in each compartment.

5　The individual crown boards are put into place and the roof replaced.

6　Calculate the time after which the virgin queens will have emerged, mated and should be laying eggs.

7　After this time check the compartments to determine if the queens have emerged, mated and are laying eggs. When each queen has produced a 5cm diameter patch of worker brood assess the quality of the queen and her brood by checking for a regular pattern of laying.

8　These small nuclei can be used to re-queen other colonies by uniting them. See Section 9 -Uniting bees.

Swarm management using the Taranov board

The Taranov Board is a plywood sheet with a supporting frame that provides a sloping surface from the ground to the hive entrance. A simple board wedged below the hive entrance and the ground will suffice as long as there is sufficient space (i.e. no long vegetation) for a cluster of bees to gather underneath the board.

The technique relies on the queen having been clipped and unable to fly. On emerging during swarming she will fall to the bottom of the slope and move away from the light to seek shelter under the board. Bees then cluster around her and the queen and cluster can be recovered and hived. Rather than try to return them to the colony it is best to carry out the artificial swarm technique as described elsewhere.

De-queening as a method of swarm management

The removal of the queen and the subsequent re-queening of the colony through the production of queen cells by the de-queened colony is a method of swarm prevention and control.

It can be used to ensure a break in the production of brood that can be advantageous in disrupting the life cycle and subsequent re-infection of brood cells by Varroa. The absence of brood means that the Varroa mites have to live on the external body surfaces of the adult bees where they are vulnerable to chemical control methods.

The disadvantage is that, unless the de-queened colony contains suitable queen cells for selection at the time of de-queening, the new queens will be raised from emergency queen cells and thus probably not reared and mated under the conditions necessary to achieve well developed and properly mated queens. Also the break in the brood can set back the development of the colony and risk missing a potential nectar flow.

CATCHING AND HIVING SWARMS

Swarms of bees can become available from a number of sources including those from your own bees, swarms that arrive in your apiary from elsewhere, and those notified to you by others. Unless the swarm has been out of its hive for a few days, especially during cold wet weather, bees in swarms are quite docile. It is always advisable however, to approach all swarms wearing your veil, as you may not know the swarm's temperament until you start to work with it.

The tips of branches, tops of fence posts, hedges, roof spaces, chimneys, wall cavities, even post boxes are all suitable places for swarms to alight and cluster. Some of these, especially those in public or third party locations may be difficult to deal with. At all times you should carefully consider the risks to you, your helpers and members of the public. Be aware of your insurance situation in relation to personal injury and damage to the property of a third party. Membership of your beekeeping association may already provide such insurance cover but you should check this. Remember you are not obliged to deal with another beekeeper's bees, feral or wild bees or swarms reported to you by third parties.

Collecting and hiving swarms requires patience and technique, but we never cease to enjoy watching a swarm enter a hive through the entrance after we have collected it. The key to successful swarm collection is to get the queen into the swarm collection container. Swarms can be collected in straw skeps, cardboard boxes, nucleus or brood boxes with floors. Plastic buckets with a capacity of 25l and snap fit lids with wire gauze inserted into the lid to provide ventilation also make a good means of catching and holding swarms and it is worth constructing one.

A good swarm kit consists of:

- Personal protective equipment.
- A smoker, fuel and matches.
- A container in which to collect the swarm.
- A bee brush or goose wing.
- A super frame containing some honey stores.
- Straps or tape.
- A Snelgrove board.
- Secateurs for removing branches and other vegetation.

Swarms will usually move upwards into the darkness provided by a container placed above it. This behaviour can be used to your advantage if the swarm is on a post or in a hedge or even on the ground and it is possible to support your container over it. In these circumstances smoke should be used sparingly, otherwise it can cause the swarm to take flight and be lost.

When swarms enter air bricks or chimneys the best strategy is to try to smoke them out, so they will settle nearby, hopefully where they can be caught more easily. In such circumstances once you have removed the bees, close the hole from which they emerged or where they might have gained access just for the summer. This is to prevent other swarms from picking up the scent of the scout bees of the first swarm, and using the site themselves.

If the swarm cluster is hanging freely for example from a branch, position the collecting container immediately under it and, if possible, raise the container so the cluster is inside it. Dislodge the cluster with a sharp upward movement of the container or a downward jerk on the branch; in some cases you may have to cut the branch to get the swarm into the container. Take care if you have to use a ladder. If at the first attempt you catch many of bees in the container, quickly lower it to the ground and up end it over a white cloth or sack, this makes it easier to see the bees and to observe if the queen is with them. Place a super frame so that one of the lugs of the frame supports the container to encourage the flying bees to join the swarm.

Catching the swarm.

Inevitably there will be a number of bees in the air, but if you have caught the queen at the first attempt, you will see worker bees around the edge of the container with their abdomens facing outwards exposing their Nasonov glands. These glands produce attractant pheromones that guide the rest of the bees to the queen. Do not worry if some of the airborne bees seem to cluster again, if the queen is in the container they will be attracted to her or return to the original colony.

If you have not caught the queen the bees will leave the container and cluster again with the remainder of the swarm. Wait until the swarm is settled before repeating the attempt to catch it. When it becomes obvious that the queen is present and the bees are entering the container you can either wait until all the bees are inside or leave them to it and come back later in the afternoon or early evening when the majority of the bees will be in the container.

Look at the entrances of your hives if you suspect that it is your bees that have swarmed. If a hive entrance has a higher amount of activity compared to other hives in the apiary it is likely that the swarm has come from that hive.

When you move the swarm ensure that the collection container is closed but ventilated. Take particular care if you are transporting it in the passenger compartment of your car. In the apiary the swarm is then hived by dislodging it from the container onto a white sheet supported by a board to enable the bees to walk into the entrance of the hive that has been prepared for them. When all the bees have entered, remove the sheet and feed the swarm with at least 3 litres of sugar syrup (1:1sugar to water by weight).

An alternative method is to place an empty brood box on top of the prepared hive without its roof and crown board and empty the bees directly from the container onto the combs. Then remove the empty brood box and replace the crown board and roof. Feed the swarm with at least 3 litres of sugar syrup. If the weather is poor and forage scarce you may need to feed them again. Take care not to spill syrup in the apiary which can promote robbing. See Section 9 - Robbing bees. Swarms will produce excellent straight brood combs drawn from foundation.

Not all swarms are headed by mature queens. Sometimes the swarm will contain several virgin queens and even though you think you have successfully hived the swarm it will abscond. A good technique to help increase the success rate of keeping swarms once they have been hived is to place a queen excluder between the floor and the bottom of the brood chamber to prevent her from flying from the hive. The technique is not foolproof but it may help to keep the queen(s) in the hive until you have time to establish their status.

Aftercare of swarms

Some beekeepers have a separate apiary location specifically for swarms they collect in case they are diseased. Unless you can be certain of the origin of the swarm, you will not know its disease status or its temperament. It is advisable to treat the swarm with Apistan® or Bayvoral®, Apiguard® or oxalic acid solution to rid them of any Varroa mites that are present on the bees. It is during the time when the mites are on the adult bees that they are most vulnerable to control by the acaricide.

As the swarm colony develops and brood is raised carry out checks for brood diseases. When working with the swarm colony monitor its temperament and how the bees move on the comb. If they rush wildly about from one side of the comb to another it is more difficult to make satisfactory inspections, and re-queening with a queen whose progeny do not carry this trait is advisable.

It is not good practice to use swarms to augment existing stocks or to build up weaker stocks without assessing their disease status as disease can be introduced in this way.

SOME HISTORICAL NOTES ON SWARM CONTROL

Like many other crafts and occupations, the names of inventors and discoverers are often given to techniques and equipment. Beekeeping is no exception to this. So for the sake of completeness some short descriptions of methods of swarm control named after their inventors are included. Books on beekeeping often refer to the technique by the originator's name rather than a more helpful descriptor.

Pagden method
(this is similar to the typical artificial swarm method used today)

This method can be used either if a colony is found to be developing queen cells or if you want to prevent it swarming. A new hive is placed beside the colony and its brood chamber filled with either frames of foundation or drawn comb. The parent colony is then examined until the queen is found and the comb on which she is found is placed in the new hive's brood box; destroy any queen cells present on it. Preferably the comb should contain only sealed brood and if it is available the queen should be transferred onto it. The hive now containing the queen and foundation or drawn comb is assembled to form the artificial swarm. The queen will start laying eggs and the colony will be augmented by foraging bees returning to the original site.

The frames containing brood and their adhering bees are contained in the reassembled parent colony hive and this is moved to another site in the apiary. A suitable queen cell is selected and allowed to develop into a virgin queen, emerge and mate. Any remaining queen cells are destroyed or used elsewhere. This means there is a long break in brood rearing which can be avoided by introducing a mated queen into the colony.

Advantages	Disadvantages
• Usually used at a time when the colony is rapidly expanding • On average the bees are young • There is no reduction in laying • Virtually 100% successful • Provides a new queen for the year • Number of colonies doubled • Bees and honey are not lost	• Additional hive parts required • Good flying weather is necessary • Bees may abscond • There is a break in brood rearing unless a mated queen is introduced (requires finding queen(s) • Does not always enable bees to take advantage of the main nectar flow • No genetic selection

Demaree method

This method relies on the principle of separating the queen from the brood in the colony.

The queen on a comb of brood is removed from the hive and placed in a second brood box filled with empty combs. The hive is reassembled on the same site by placing the brood box containing the queen on the floor, inserting a queen excluder above it, and adding the parent brood chamber including the original brood, cover board and roof.

The bees, except the queen, have full access to all the combs. The upper brood box should be examined for queen cells which can either be destroyed or utilised elsewhere. The colony generally settles down and makes no further attempt to swarm and the original parent brood chamber and combs can be removed once all the brood has emerged and the bees in the upper brood box have been cleared.

Advantages	Disadvantages
• Seldom necessary to use the technique more than once per season • Easily carried out on the same hive stand • It spreads the brood and the adult bees reducing congestion • Bees and honey are not lost	• Effort has to be spent by the colony getting additional brood combs drawn out • Requires finding the queen • Unlikely to be 100% reliable • Requires careful searching for queen cells

Snelgrove method

The basic method involves the use of a Snelgrove Board which is a cover board with opening and closing wedges set in the upper and lower edges of the board, and a central opening covered with fine zinc mesh. The method is the same as that of Demaree, except that the Snelgrove Board is placed on the supers. For the first 5 days after setting up the colony only one upper wedge is opened in the Snelgrove Board, which acts as the floor of the parent colony, to allow some flying bees to emerge and rejoin the swarm below, using the entrance in the bottom brood box containing the swarm and the original queen. After 5 days the top wedge is closed and the corresponding wedge below is opened, thereby allowing foraging bees to return and join the bees in the supers. At the same time the upper wedge on the opposite side of the Snelgrove Board is opened. After another five days this wedge is closed and the corresponding lower wedge is opened. In this way maturing bees are drafted into the supers on a steady basis. When all the brood has emerged from the parent colony the Snelgrove Board can be removed and the brood box used as a honey store. The queen cells present in the parent colony can either be destroyed or used to increase colonies.

Advantages	*Disadvantages*
• It only needs to be done once per season • Can be used to produce new mated queens • Can have 2 queens laying at the same time • Bees and honey are not lost	• Time consuming • Requires frequent visits to the hive to adjust the entrances • Requires additional equipment • Requires finding the queen

SECTION 8

KEEPING HONEY BEES HEALTHY THROUGH THE YEAR

This section describes a typical year of practical beekeeping in the British Isles, bearing in mind that no two years are ever the same and conditions will vary with geographical location. It demonstrates the ways in which the theoretical sections of the book are put into practice in the apiary.

Climate change

In future years it is probable that climate change will influence the behaviour of honey bees which in turn may change the current beekeeping calendar. Agricultural practices will continue to change and existing forage sources may diminish and be replaced by others, either more or less suitable. If less suitable this will limit the surplus honey beekeepers can harvest without compromising the survival of the colony. If winters become warmer colonies may continue through the year without a brood-free period, exacerbating the possibility of brood diseases. More changeable and extreme weather during the period in which the queen is making her mating flights may lead to poorly mated queens unable to provide a good supply of fertilised eggs.

Getting started

It is essential to record the findings from inspections, health checks and other beekeeping activities for each colony. Record keeping is also interesting and useful for comparing beekeeping activities from year to year. Only with good records can you plan what needs to be done with any degree of certainty. It is good practice to start keeping records from the time you begin beekeeping. If you are a new beekeeper now is the time to think about establishing a system to suit your needs, and the practicalities of how you will use it. Ask other beekeepers how they keep records and learn from their experiences. My method is to take a notebook and pencil into the apiary, having first checked my previous notes for each colony. I write the notes using shorthand, such as QR for queen-right, and it helps if each hive is numbered individually. These notes can be transferred onto index cards or alternatively, set up a computer based recording system. The British Beekeepers Association (BBKA) has a leaflet on keeping hive records. For more details see Section 9–Keeping Records.

In winter it feels too cold and damp to spend time in the honey house or workshop cleaning and repairing equipment however, suddenly the new beekeeping season begins and there have been times when I wished I had been more diligent in cleaning and preparing my bee equipment in preparation for the new season. Whenever you decide to spring clean have a check list of what needs to be done.

- Carefully examine your stored brood combs; can you see light through them? If not, the old comb should be removed, wrapped in newspaper and discarded in the household waste, old or unwanted frames can be discarded in the same way but preferably they should be burnt.
- Scrape and scorch hive boxes with a blowtorch or scrape frames and sterilise them in hot washing soda solution (see pack details) to reduce the incidence of disease in your colonies. Allow the frames to dry thoroughly before you store them or insert new foundation.
- Clean queen excluders, lay them on a flat supportive surface for cleaning and take care with the zinc/metal slot types, bending or distorting the metal, or enlarging the slots may prevent them from forming an effective barrier to the queen, especially young queens before they have matured to their full size.
- Examine the springs in the Porter Bee Escapes and remove propolis on the springs and inside the escapes. Washing soda solution or denatured alcohol can be used for this.

THE SEASONS

The times shown are indicative only and will vary significantly in some years and depend on geographical locations and weather conditions.

Late winter–January until late February

Remember:

• monitor for Varroa and treat if necessary.

February can often bring the worst of the winter weather with snow, rain, cold temperatures and strong winds, at the time when the colony is beginning to expand. Beekeepers need to be aware of weather patterns by watching or listening to weather reports and forecasts; the internet is a good source of mid to long term weather data. This information will help you to plan your beekeeping activities.

During the winter months honey bees cluster on the combs, the colder the weather the tighter the cluster. They are capable of maintaining their cluster temperature by progressively moving as a cluster across the combs as they consume their food stores. In mild winters, or in the parts of the British Isles with milder climates, the queens may have already started laying eggs in late winter and young brood may be developing. Do not be tempted to see if the colony is still alive by taking off the roof and having a look, or kicking the side of the hive to see if any bees fly out. On fine days the bees will fly and defecate on the wing; also some bees will fly out to collect water to dilute the honey stores so they can be utilised. On such days take the opportunity to observe from which colonies bees are flying. Look carefully at the alighting board, and on the ground immediately below the hive entrance. A few small particles of wax indicate that the bees are progressively uncapping their stores, but the presence of larger pieces may mean that there is a mouse in the hive. There may also be a few dead bees present, but unless there are many this should be no cause for concern.

Books and beekeepers will often advise you to put a quilt on the cover board under the roof to maintain the temperature in the hive. Quilts or other covers can obstruct the flow of air through the hive, condensation will form on combs and internal walls and damp is detrimental to bee health. It also promotes mould growth and honey fermentation in the comb. To avoid this always ensure there is good ventilation and movement of air through the hive.

Good practice during the winter months is not to cover the holes in the crown board, or put any material on top of the brood frames that will restrict the air flow through the colony, also ensure that the vents in the roof are not clogged. This will enable the cluster to regulate its temperature and relative humidity. If you have fitted mouse guards check that the holes are not blocked by dead bees or other debris such as leaves. Blocked holes will restrict the air flow and prevent bees leaving the hive for defecation flights. After snow fall check and remove snow blocking the hive entrances.

Make sure the hives are off the ground to allow ventilation under the floor, this helps to keep the floor dry. The hives should slope very slightly forward to prevent rain or condensation from draining back inside the hive. Check for signs of woodpecker damage. These birds will often attempt to drill through into the brood boxes especially in the hand holds of the older styles of boxes.

From the time when the bees first cluster in late autumn / early winter to the end of February the colony will consume between 2.27 – 4.55kg (5-10 lb) of stores. From the end of February to the end of April an average colony will consume 4.55 – 9.1kg (10-20 lb), depending on the temperature and the rate of brood rearing. In mild winters bees may consume their stores very quickly because of increased activity, and it is important to be aware of the food status of colonies. One way to do this is to use a spring balance and lift opposite sides of the colony from under the floor noting the weight on each side. Add the two weights together to give an approximation of the hive weight. Do this at the start of winter and then monthly and record the weight loss. Another way is to lift (heft) the sides of the hive and estimate the weight by comparing it with a similarly configured hive without bees and stores. With practice this can be a very useful technique.

If your colonies need emergency feeding the best solution is to use a super containing some frames of uncapped honey placed on top of the brood box. When harvesting your honey always retain some supers for this purpose. Alternatively place a block of fondant on top of the cover board, or soak a 1kg bag of granulated sugar in water, allow excess water to drain off, cut open the bag so the bees can access the sugar, and place it cut side down adjacent to or over one of the holes in the cover board.

Bees need water to mix with the stored honey so a conveniently close source of clean water will enable forager bees to collect water and bring it back to the colonies quickly to minimise exposure to the cold. A shallow tray filled with gravel and clean water is a good method, it enables the bees to access the water without drowning. Remember to check the container and replenish the water when necessary.

Early spring–March

Remember:

- monitor for Varroa and treat if necessary.
- remove the mouse guards if you use them and replace them with entrance blocks.

By early March colonies should start to expand with increasing amounts of brood being reared, the sight of worker bees returning to the hive with pollen loads is a good indicator of this.

Brood rearing puts heavy demands on the colony's food reserves, and early spring is a time when stores can run very low. In extreme cases the colony will either die and on inspection you will find the heads of the starving workers deep down in the bottom of cells trying to access the last of the honey stores. A sign of starvation is the presence of larvae on the alighting board, although you may not see them because they are favourite morsels for birds such as tits, consequently you may not be aware that the colony is starving.

Remember it takes over three weeks after the day the egg is laid before an adult worker emerges from her cell and an additional and variable amount of time before she will forage. Reduced egg laying or starving workers cannot produce the number of bees needed for the foraging force, and such colonies will be small and unable to exploit early crops such as oil-seed rape. At best the colony will build up on the early crops rather than providing a surplus of honey for your harvest. Check the hive weights by hefting them regularly and if necessary, feed the colonies with sugar syrup.

This is the time to carry out a quick inspection to check colony survival and condition. See Section 5 – The Essentials of Bee Husbandry.

Do not be tempted to carry out your first full inspection before the air temperature is above 14°C. In some areas such as the southwest this may be as early as February, but elsewhere it will be March or even April. Remember the less you disturb the colony the quicker it can carry on with its normal activities and take advantage of the better weather. Details for carrying out a full inspection are contained in Section 5- The Essentials of Bee Husbandry

Mid spring–April

Remember:
- monitor for Nosema and treat if necessary.
- start checking for Foul Brood and continue periodically until September and if you think your bees are infected inform the bee health authorities.

As the weather improves, abundant forage can bring on colonies in leaps and bounds, but it is also the time when the colony is most vulnerable to bad weather. Many of the workers bees will have over- wintered and will be exhausted by the demands placed on them for surviving the winter, rearing early brood and early foraging so they will be nearing the end of their lives.

Young worker bees take several weeks from when the egg is laid until they become foraging bees and if the level of stores in the colony is low this is a potentially risky time when the colony could run out of food. If the colony deploys too many bees in foraging there is a real risk that the brood will not be adequately nursed, and if there is too much brood for the nurse bees to nurture, it may become undernourished, chilled and die. Figure 1 shows a typical annual population curve of brood and adult bees in a colony that has not swarmed. It indicates that there is a period from March until well into May when the brood outnumbers the adult bees.

Factors such as speed of colony build-up influence the choice of strain of bee you keep. A strain with too rapid a build-up in a cold spring when no forage is available may starve unless you feed it. Large colonies of bees confined in their hives because of bad weather may initiate swarming preparations and during the first period of fine weather in late April / early May some of the colonies may swarm.

By mid April the focus of colony inspections should include planning for queen replacement and swarm management and control. You should also plan ahead if you want to increase colony numbers, or develop the strength of your colonies. It is helpful to mark the queen so you can locate her quickly on the comb. You may also want to clip her wings as part of your swarm control regime, but clipping the queen's wings does not prevent swarming altogether.

There are actions the beekeeper can take to delay the onset of swarming; the principal one is the provision of adequate space for the expanding colony. The aim is to provide comb space where the worker bees can unload and process the increasing amounts of nectar gathered. If there are no supers the bees will place the nectar in the brood comb; not only does this significantly reduce the amount of brood the colony can raise, but it also means more congestion in the hive. This can be reduced by adding a super of ready drawn comb as the bees prefer this to a super of foundation. Otherwise they may ignore the foundation and continue to place and process the nectar in the brood combs. If the colony is expanding rapidly, a second super placed under the one with drawn comb can be of foundation only.

Each time you carry out an inspection identify those brood combs that need changing, usually at least every two to three years. The regular changing of combs is now considered to be an essential part of an integrated bee health management regime and will help control the incidence and severity of brood diseases such as Chalkbrood, EFB, AFB and Nosema.

In many parts of the country autumn sown oil-seed rape will be flowering, as will top and soft fruits and dandelions which are an important source of nectar and pollen. By this time colonies should be expanding quickly and if you have a colony where this is not the case and assuming that the food reserves are adequate, in most cases the cause is due to either queen failure or to Nosema.

Late spring–May

Remember:
- monitor for Varroa and treat if necessary.

Even in late spring the weather can continue to be variable with long periods when the bees cannot forage and when the air temperature is too low for the forage plants to secrete nectar or for flowers to open and yield pollen so there is still the chance of starvation and you should be prepared to feed.

Normally this is the time when the main phase of swarming takes place, and the factors that promote the onset of swarming will already be present in the colonies. The presence of drone brood and young drones indicate that the colony will soon begin to rear queen cells if it is going to swarm. They also enable you to calculate the approximate time when you can watch for swarms. Drones are sexually mature around 37-40 days after the egg is laid, and the earliest the colony is likely to rear queen cells is around day 4 /5 after the drones begin to emerge from their cells, but like all things in beekeeping there is no hard and fast rule. There are techniques the beekeeper can use to help reduce the possibility of a colony swarming or to manage them more to his or her convenience, see Section 7- The Management of Swarming.

It is always good practice to work with the natural instincts of the bees and this is particularly so in the case of swarming. In my experience queens raised from swarm cells are usually the best as they are likely to have been produced during a period of good forage, in a colony with plenty of nurse bees. It is advisable only to select the queens from colonies that produce relatively few queen cells (< 10) as they are less likely to be a 'swarmy' strain.

I always assume that every colony is going to swarm at some time and the trick is to manage your bees so that you can artificially swarm the colony at a time to suit you and the bees, and you retain control of the situation. Once this is done, you can decide what to do with the artificial swarm, the original colony and its queen cells.

When the weather is warm and your bees are flying to the nectar flow on a crop like oil-seed rape, it is surprising how quickly the supers can be filled. Be prepared with spare supers, preferably of drawn comb, to put onto the colonies. However, if the frames are only fitted with foundation, place them in a super on top of the brood box so the warmth from the brood nest can help the bees to draw out the foundation. It is better to give too much rather than too little super capacity. Do not wait until the bees have filled all the combs and are drawing comb on the crown board.

FIGURE 1. Annual population cycle of adults and brood in a honey bee colony.

■ Brood

■ Adults

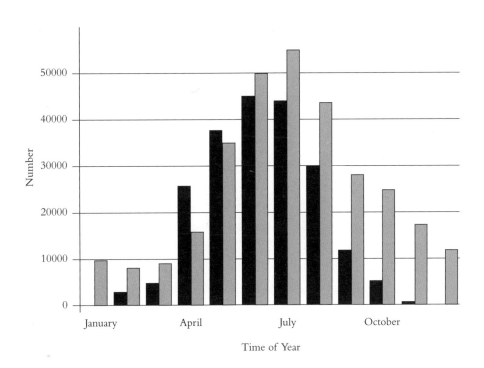

Foraging on oil-seed rape

Colonies of bees placed within or very near to crops of oil-seed rape usually perform better in terms of honey production compared to colonies whose bees have to fly some distance to the rape. If you place your bees in the rape crop make sure the farmer / contractor knows exactly where they are located so they do not get damaged by machinery or by spraying with pesticide products.

When bees forage on oil-seed rape the beekeeper has to consider how to deal with the honey which will granulate quickly in the comb and be impossible to extract using a mechanical extractor. If you extract the honey before it is capped there is the risk that its moisture content will be too high and thus unsuitable for sale, or it could ferment in the storage containers. To test this take out sample combs and hold them horizontally over the top of the super, if the honey runs out it has too high a moisture content and needs further evaporation by the bees, so replace it in the super.

When the honey is ready the usual way of clearing the supers of bees is to use clearer boards. These are cover boards with one way bee valves so that once the bees go down through them into the box below the clearer board, they cannot return into the super. It is also possible to take out selected individual combs of honey and use a bee brush or goose wing to dislodge the bees from the combs before placing them in another super or box with a floor and cover to prevent the bees regaining access to them. The combs can then be taken away for immediate extraction. The more complete the extraction of honey from the comb the more you reduce the potential for residual oil-seed rape honey which will initiate the granulation of subsequent honey crops.

An alternative approach is to leave the combs on the colony until all combs are sealed, remove them and store them in a cool dry place until you are ready to process them in a container warmed by a water bath. Some beekeepers scrape the oil-seed rape honey and comb down to the foundation placing it directly into sealable containers for further processing. The disadvantage of this technique is the loss of drawn comb, however the wax can be recovered.

During your inspections check for any signs of unhealthy bees or larvae and match up the signs with those described in Section 3 - Honey Bee Diseases and Conditions, then carry out the suggested actions. It is during this period that diseases such as European Foul Brood (EFB) become noticeable.

As the numbers of drones build up so does the potential for the Varroa mite population to increase so monitoring should continue, also look for bees, both workers and drones, with deformed wings and other abnormalities. Remember that Varroa mites act as vectors for bee viruses which may be present in healthy colonies with low Varroa mite populations. When the mite levels increase their feeding on the bee's haemolymph, viral transfer infects and debilitates an increasing number of bees.

Early summer – late May-June

This can be a good period for forage, especially if your bees are working crops such as field beans. The colonies will be reaching their most populous and a strong colony will require at least 3 supers to prevent it becoming congested.

Swarm management and control continues to be a key part of beekeeping during this period. It is also a good time to raise new queens or through using the queens reared as a result of your swarm control and colony splitting techniques. Continue to monitor the colonies that have either not swarmed or from which you have not made any artificial swarms and provide space for the bees to place and process the nectar they collect.

Mid summer – late June-July

Often this is a period when there can be a shortage of forage at a time when colonies contain their maximum number of bees. Also in some years July can be a wet month, making the bees unwilling to leave the hive to forage. It is important to keep hefting the hives to check that there are adequate food reserves, especially if you have been harvesting honey in the previous weeks.

Colonies containing the season's new queens should be carefully monitored to check if they are laying eggs and producing both worker and drone larvae. This confirms that they have been mated successfully. The period of time between the emergence of the virgin queens and the first sighting of eggs varies considerably according to the season and the weather conditions. In general you should wait at least 14 days after you believe the virgin queen has emerged to take her mating flights before going into the colony to see if she is laying eggs. If you look too early you can upset or injure her or even prompt the colony to ball and kill her, alternatively she may abscond with a swarm.

It is always worth noting and recording the weather during this period, as this will give some indication of whether the virgin queen will have been able to fly to mate. A new queen that has not mated can remain receptive to mating for 3, possibly 4 weeks, after which she cannot mate successfully and she becomes a drone layer.

Working bees on borage

In some parts of the British Isles borage (also known as starflower) is a crop from which substantial yields of honey can be harvested, mainly in July. In warm weather the rate of the nectar flow can be high and it is not uncommon for a strong colony to fill a super of drawn comb in a day. If your bees are working borage you need to provide them with plenty of super capacity, otherwise they will build wild comb in any available space and fill it with honey. Trying to deal with such situations is a sticky business and often results in many dead bees drowned in the honey flowing from broken wild comb. As well as the honey crop, borage is also very useful for drawing out new perfectly straight brood combs from foundation. When filled with honey these frames act like supers and make valuable stores for feeding over-wintering colonies, or using for emergency feeding.

Late summer – late July to early September

Remember:
- monitor for Varroa and treat if necessary.
- monitor your colonies for Nosema and in any case treat with Fumidil®B.

The colony's population starts to diminish during this period. If the colony has no need to replace the queen the drones will be progressively herded outside the hive, denied access to food and subsequently they will die; sometimes they can be found clustering under the alighting board. If the workers allow drones to remain into the late autumn it may be an indication that the colony will supersede.

For many beekeepers this period is the make or break part of the season, especially if their main crops are Ling heather (*Calluna vulgaris*), heaths (*Erica tetralix and E. cinerea*), bramble and willowherbs. 'Heather-going' is a long established tradition practised by beekeepers, and in order for colonies to gain the maximum from the forage available careful preparation of the colonies is necessary. For some beekeepers the fact that colonies collect their winter stores on the heather is sufficient justification for taking hives to the moors.

In parts of the British Isles borage will continue to flower until it is swathed, even then there is usually some re-growth and borage will continue to flower until the first frosts kill it off. Mild autumns can ensure that bees have additional forage sources such as ivy, Himalayan balsam, red clover and knapweed. These can bulk up the stores, so have at least one extra super on each colony to take advantage of late nectar flows.

Wherever possible ensure that the colonies are able to forage on pollen sources all through the active season to ensure that they are not short of pollen, which is essential for the development of new bees, especially as at this time of year colonies are producing bees that will overwinter until the following spring. These bees are crucial for the over-wintering to succeed and to kick-start the brood rearing process. Over-wintering bees are physiologically different to summer bees, they develop fat bodies which are reservoirs of proteins in the abdomen, such as vitellogenin and other substances including glycogen. Fat bodies enable bees to produce brood food in their hypopharyngeal glands in the late winter and early spring. During cold snaps contraction of the bee cluster may prevent individual bees accessing the pollen required for feeding the larvae, and if this happens the stores in the fat bodies are used.

By the end of September beekeepers will often have several small colonies in the apiary, they may have been casts or small nuclei that have not developed, and if left they are unlikely to survive the winter. In a good year there may be too many colonies for the beekeeper's needs. The answer in these situations is to unite them.

The main task at this time is preparing your colonies for successful over-wintering. In recent years winters have become noticeably milder (an ambient temperature of greater than 5°C) and wetter in most parts of the British Isles. This means that colonies tend to use up their food reserves more quickly because the cluster is more active. Also there is an increasing tendency for brood rearing to take place throughout the winter, especially in milder parts of the country. Ensure that the colony is well provisioned with food to last well into the following spring, monitor the rate of depletion of the stores and feed when necessary. It is vital that colonies go into the autumn and winter with young healthy bees that have well developed fat bodies and low mite populations. Both these factors are under the control of the beekeeper.

Heather Going

Heather going is best suited to single walled hives which are easily secured, will pack well and are easy to handle. Sometimes at lower elevations on the moors there will be willowherb, bell heather and wood sage and this will give a heather blend honey. Usually the higher the elevation of the moor the lower the air temperature, and stronger winds require you to place the hives in a sheltered position. The best heather for nectar secretion is flowering young heather and this is to be found on regularly burnt moors.

Ideally choose a site where there is >40 ha (100 acres) of heather moorland and has easy vehicle access in all weathers. The hives should be set out in pairs or groups facing different directions. Do not set up the hives in lines otherwise the bees will drift. Keep the bees and their flight lines away from shooting butts and places where the public or shooting parties may take their lunch or rest.

Colonies taken to the moors need to be really strong and contain brood at all stages, a young and vigorous queen and adequate stores. If the hives have two supers place one or two drawn combs in the lower super. Double brood colonies should be reduced to one box by confining the queen in the lower box 3-4 weeks before moving to the heather.

During transport the hive entrances should be blocked so that they are completely dark. Provide ventilation by using a wire mesh travelling screen, mesh floors will also greatly assist in the ventilation. Take some Plasticine® and sticky tape to plug holes in an emergency.

Moving colonies can aggravate Nosema and it is recommended that bees that have been to the moors are fed with around 4-5kg (10 lbs) of sugar syrup (2 parts to 1 part water) containing Fumidil®B on their return. The colonies may also require Varroa treatment. Any unused brood comb and hive boxes should also be fumigated with acetic acid vapour.

If possible it is recommended that you clear the honey supers and transport them separately from the colonies as the unwired combs are fragile.

Autumn – Late September to October

This is a time when late forage can make a welcome addition to the overwinter stores. If you are leaving a super on the colony over the winter remember to remove the queen excluder so the queen does not get left beneath it if the cluster moves above it. If you clear the super and extract the honey remember to check its moisture content. If it is above 20%, mix it with some honey that has a lower moisture content so that the blended moisture content is below 20% and suitable for storage and sale. For heather honey the moisture content should not exceed 23%.

I usually overwinter the majority of my colonies on mesh floors, either in double brood boxes or a brood and a half. In the latter case the super is placed below the brood box and above the mesh floor. This extra space helps the colony to utilise the food reserves more effectively and not become isolated on the tops of the frames in the brood box, away from their food; it also gives adequate clustering space. In double brood colonies there are usually several good food combs in the upper box often containing valuable pollen. I always keep back a few supers of sealed honey for feeding bees if required, using honey from crops such as borage or balsam that are unlikely to granulate in the comb. Combs containing oil-seed rape honey which granulates in the comb should not be kept for feeding, so if you do not have supers of liquid honey use refined sugar-based stores for over-wintering.

In some years it is quite common for the queen to stop laying eggs in the early autumn, especially if the weather is cold and there is little forage available. If you inspect a colony at this time and find no eggs, do not immediately assume there is a problem with the queen because often a colony will supersede its queen late in the summer and the new queen may not yet be laying eggs. When inspecting your bees note if there are still drones present in the colony, and whether there are supersedure queen cells on the face of the comb. The presence of supersedure cells does not necessarily mean that the queen is damaged or deficient as the routine supersedure of an undamaged queen in some strains is thought to be a genetic trait. If you find supersedure cells let them develop, allow the virgin queen to mate and begin to lay eggs, do not remove the original queen as she may be laying eggs whilst the virgin queen matures. It is the workers that will eventually decide which queen is to survive, unless you decide otherwise. The presence of two coexisting queens, confirms that supersedure has taken place. Autumn is also a good time to re-queen colonies and provided certain precautions are taken, this can be carried out successfully at this time.

A question often asked is whether to store your honey combs in a dry or wet condition (with the remains of the honey not spun out during its extraction). A dry condition is achieved when after extraction the supers are placed above the crown board so the bees can clean them up and take down the honey for their stores. If the combs are left wet the remaining honey will absorb moisture during the winter and may begin to ferment. If there are combs containing a lot of pollen and you are over-wintering the colony on a brood and a half, or with a honey super, place these combs towards the centre of the super below the brood chamber so the bees can process the pollen for storage over the winter.

As winter approaches ensure that your bees have enough stores and make final preparations. Hives are very attractive places for mice to overwinter, with a source of food at hand. When the bees are tightly clustered, mice can gain entry, build nests, consume food and soon demoralise the bees leading to colony dwindling. One way of keeping mice out is to use mouse guards. My preference is to use narrow entrance blocks all through the year, these have only sufficient headroom for a bee to enter the colony and will satisfactorily exclude mice.

Ventilation is important and should be achieved by leaving the feed holes in the crown board open and making sure the vents in the roof are clear of debris. The hive should be kept well off the ground. The roofs should be weighted down with a brick or a stone, and the hive should be tilted slightly forward so that any water blown into the hive entrance will run out. If the colony is being over-wintered on a mesh floor this is not an issue.

Winter – November to January

This is the period when the British Isles usually has its first spell of really cold weather, often in November, followed by a second spell in February. During this period the bees will be clustered, the tightness of the cluster depends on the ambient temperature; the colder it is the tighter they will cluster. On warmer days in late winter bees can be seen leaving the hive on defecation flights and, if there are early bulbs flowering, a welcome sight is to see pollen being taken into the hive, a sure sign that there is brood being reared. Bees can also be seen visiting water sources.

Hives should be regularly hefted and a mental note of the hive weight made. If a hive seems much lighter than when previously checked it may be necessary to give it some emergency feed. Regularly check the hives for damage and keep the entrances clear. Expect to see some dead bees behind the mouse guards or on the alighting boards.

Feeding during the winter months can be difficult because the bees may not take down the syrup provided; even if they do there is a risk of fermentation in the comb if they cannot reduce the moisture content of the syrup. Good practice is to provide a slab of fondant that can be placed across one of the holes in the crown board to allow bees to access it. In emergency situations place it on the frames, directly over the cluster.

Time spent reflecting on the season and considering the value of the techniques you have used and whether you can improve on them will assist in planning for the coming season. It is always tempting to think that merely increasing the number of hives will automatically mean more honey. Fewer, well managed and disease-free colonies and access to good forage throughout the season will always yield more honey and more pleasure to the beekeeper than a larger number of colonies that are less well cared for.

SECTION 9

USEFUL TECHNIQUES

A - DEALING WITH DIFFICULTIES
AGGRESSIVE COLONIES

Handling large aggressive colonies can be a daunting prospect even for experienced beekeepers so ensure your bee suit is bee proof, tell someone where you are in case you are badly stung and suffer a reaction, have some antihistamine tablets with you and where possible have another competent beekeeper present to help you. Some strains of bees are habitually aggressive, resenting any attempts by the beekeeper to inspect them; however the aggressive behaviour may be due to causes other than beekeepers activities, for example:

- The colony is queen-less or the queen has been superseded.
- The beekeeper has introduced one or more queens in the previous season and the subsequent crossing with local strains the following year has led to a new queen with inherited aggressive tendencies. Before buying in queens, carefully consider whether or not the second generation crosses could result in a colony with aggressive traits. Ask your queen supplier about this.
- Robbing by other bees or wasps or being disturbed by large animals or vandals; this may be temporary.
- Abrupt cessation of the nectar flow.
- Bad weather e.g. thunder.
- The colonies may have experienced spray damage.

Using too much smoke in trying to subdue an aggressive colony usually results in the beekeeper retreating from the apiary. A hand held water spray containing clean water is often very effective in cooling the bees' aggressive behaviour. The situation can be further improved by separating as many as possible of the flying bees from those nursing the brood. This is achieved by sealing the entrance to the colony and moving it, preferably on a good flying day and in early evening, to a position 10-20 m from its original site. Place a floor and brood box, fitted with a few frames of comb or foundation if available, on the original location. The supers on the original site, if present, are taken off and set aside allowing any bees to escape, they will fly back to the original colony site. Gently smoke the colony and allow time for the smoke to make the bees consume some of their honey stores making them less liable to sting the beekeeper. Put a roof on and leave them for an hour or so, preferably overnight, before beginning to examine the colony carefully and methodically to find the queen. If the combs contain a normal pattern of brood then a queen is present, decide whether she should be replaced or killed and the colony united with another colony.

Then examine the combs further, if brood is not present and there is no likelihood of there being a virgin queen or very young mated queen in the colony, take out the frames one by one and carefully inspect them for any signs of a queen, eggs or the polishing of cells which may indicate the colony has a new queen that is about to start laying. If none of these signs are seen dislodge the bees from the frames by shaking them onto the ground from where they will fly back to the original site.

If brood is present in the queen-less colony the brood box containing the brood frames from the original colony can be placed over a queen excluder above the brood box on the original site. A frame containing eggs and on which queen cells can be drawn can be inserted into the bottom brood box and the queen cells allowed to develop. Bees will move up from below to look after the brood in the top brood box until it emerges. Once all the brood has emerged the bees can be cleared and the box removed. Alternatively the best frames should be shaken to remove the bees then inserted into the bottom brood box on the original site and the colony left to raise queen cells.

COLONIES WITH LAYING WORKERS

Normal colonies may contain a percentage of worker bees with ovaries that have developed enough for them to lay eggs. This may happen if the level of queen substance pheromone 9–ODA present in the hive falls below a certain threshold or disappears due to the loss of the queen. Because these eggs are unfertilised they will develop into drones. Such drones are usually slightly smaller than those originating from eggs laid by a queen.

Typical signs of laying workers are as follows:

- No queen present in the colony.
- Only drone brood present, particularly found in worker cells.
- Eggs found adhering to the side walls of the cells.
- Several eggs in the same cell. (Note that young queens when they start laying may put more than one egg into a cell, usually located in the bottom of the cells. A normal egg laying pattern will follow after a few days).
- Scattered laying pattern.
- Eggs laid on or around larvae.
- Larvae neglected by nurse bees.
- Aggressive behaviour and even signs of workers fighting each other.

Never re-queen a colony containing laying workers as the queen will probably be killed. Never unite a colony containing laying workers with a queen-right colony as again the queen may be killed. The insertion of a frame of brood or eggs is unlikely to be successful.

Handle the colony as follows:

1. Move the entire colony a least 100 metres from the original site and place a brood box and comb on the original site.
2. Shake all the bees from the frames onto the ground.
3. Tap frames to dislodge eggs and larvae, uncap and remove all drone brood from the combs.
4. Return the frames (without bees) and boxes to the original site.
5. Add a frame of eggs and brood or a queen cell or introduce a queen in an introduction cage or unite with a queen-right colony or nucleus.
6. Feed with a dilute sugar syrup.
7. The laying workers lose their sense of orientation, they may have never have flown from the hive and are unable to find their way back the original site.

DRONE LAYING QUEENS

Typically drones will be found in normal colonies from April and May through until August. Occasionally colonies will tolerate the presence of drones beyond this period and it is not unknown for some colonies to keep a few drones through the winter. Drones are usually herded out of the colonies in late summer when, under normal circumstances, they have served their purpose. So the presence of significant numbers of drones in a colony outside these dates should prompt the beekeeper to look more closely at the queen status of the colony.

The incidence of drone laying queens seems to be increasing and an important factor may be the poor mating of queens. Wet and cold weather during the virgin queens' mating flights and the window of time when they can be successfully mated (up to 3-4 weeks after emergence) will dictate whether or not the queen receives sufficient sperm to fill her spermatheca.

The first colony inspection in spring can reveal comb with significant amounts of domed cappings, typical of drone brood, and little or no normal worker brood. This indicates that the queen has become a drone breeder. As soon as possible the queen should be found, killed and the colony united with a queen- right colony which may be a full sized or a nucleus colony. If however the queen cannot be found and, on closer inspection, drone cappings are seen on worker cells, it is likely that laying workers are present; guidance in the section on dealing with colonies with laying workers should be followed.

A mated queen can run out of sperm at any time during the brood rearing season, this leads to an increasing proportion of drone brood on the combs. Eventually no worker brood can be found. If there are supersedure cells present and they are uncapped and you can see the larvae, instead of uniting the colony, an option is to allow the new queen to develop. If you intend to unite the colony to a queen- right one then the supersedure cells should be destroyed. It is advisable to cage the queen during any colony uniting procedure.

Colonies which are suspected of having drone laying queens should be dealt with quickly as the increased number of drone cells and brood will increase the breeding potential of the Varroa mite in the colony and the impact of viruses. Where possible the colony should be monitored for a daily Varroa mite drop, and if necessary treated with an approved varroacide.

PREVENTING ROBBING

It is not uncommon for bees of one colony to rob honey from another. It may not be easily observed, but it can be obvious when large numbers of robber bees are seen gaining access to a colony which cannot defend itself. Robbing can also occur when there is an abrupt end to a nectar flow.

Robber bees are identified by the zig–zag flight pattern they exhibit in front of the colony entrance. In response the guard bees at the entrance are alerted and can be observed with their mandibles open, and their front legs raised forward. Other signs include increased activity and fighting between bees at the entrance to the hive and despite this, bees may be gaining access to the colony and leaving via a small crack or opening in the hive body. Wax debris may be seen on the alighting board.

Conditions for robbing should be avoided because, apart from the loss of stores and potential for the robbed colony to collapse, the colony being robbed may be diseased and unable to defend itself, leading to the transfer of disease to the robbers' colonies.

Wasps also rob bee honey. Later in the season when wasp colonies are reducing the amount of brood present in their nests, worker wasps start to seek out nourishment to supplement the reducing amount of sweet secretion they obtain from the wasp brood.

The following practices will help reduce the potential for robbing:

- Reduce entrance sizes, especially for nuclei or small colonies.
- Avoid spilling syrup when feeding bees.
- Feed bees late in the evening or at times when they are not flying.
- Wipe up spillages on hives, especially around the entrance.
- Arrange the colonies in the apiary to reduce the potential for drifting.
- Do not leave combs outside hives anywhere in the apiary.
- When working in colonies minimise the amount of time the boxes are open for potential robber bees to explore. Use crown boards or cover cloths to prevent access.
- Make sure the boxes fit together correctly; examine them for holes or splits through which robber bees could gain entrance. Repair the damage.
- The use of mesh floors seems to confuse wasps as they give up

trying to enter the defended entrance and try to gain access to the colony through the mesh floor, without success.

- Set up wasp traps using a weak sugar syrup.

KILLING BEES

An effective, safe and quick way is to use a strong solution of washing up liquid, 240ml (8 fluid oz) in 4.5 litres (approximately one gallon) of water. Once all the bees have returned to the hive spray the solution onto the bees using a hand held mist sprayer or alternatively a pressure sprayer with a lance, then seal up the hive.

Petrol vapour can be used but this is more hazardous. Wait until all the bees have returned to the hive in the evening, close the entrance with foam rubber, then carefully pour 140ml (approximately ¼ of a pint) of petrol over the cover board, replace the roof and prevent bees from leaving the hive. Do not have a lighted smoker, a naked flame or hot surface anywhere near the petrol or hive to be treated, also keep your mobile phone away from the site, as it is capable of igniting the petrol via microwaves. Also be aware of the potential fire risk to surrounding vegetation, especially in dry and windy conditions. Petrol vapour is heavier than air and the colony needs to be placed on a wooden floor with the entrance block closed in order to retain the vapour long enough to be effective. Leave for 24 hours, after which remove the colony from the apiary, remove the combs and dead bees and place both in a large polythene bag for disposal. If they are to be burnt first ventilate the hive to volatilise residual petrol which could ignite explosively. The other hive parts should also be well ventilated before being cleaned. First wash in water to remove any residual petrol, clean with a liquid disinfectant. After drying in the open air or a well ventilated situation scorch them with a blowtorch.

PREVENTING SPRAY DAMAGE TO HONEY BEE COLONIES

Farmers, spraying contractors, local authorities, land managers and the general public all have access to substances which if misused can damage honey bees. The incidence of spray damage has reduced significantly in recent years but vigilance is still needed. Beekeepers can help their bees, either by registering with a spray warning scheme operated by the local beekeeping association, or by ensuring that the landowner where your bees are located is aware of his or her responsibilities in relation to spraying in the vicinity of beehives or where bees may forage.

Users of pesticides are obliged by law to know whether their use of the product has the potential to harm honey bees, and if so to take mitigation measures, for example by warning the beekeeper well in advance of spraying operations. The conditions of use for any pesticides should take into account measures required to reduce the potential for damage to honey bees, for example it may define the time when application of the product should be carried out.

The simplest, but not always the most practical precaution, is to move your bees out of the area and return them when the period of residual toxicity is over. If this is not possible the bees must be confined during the period of risk. If bees are confined they require room, ventilation and water, as well as sufficient food for the period of confinement. Colonies kept on mesh floors can have their entrances blocked provided there is sufficient space to allow for air movement between the floor and the ground. Remove the roof and place a super, half full of combs, on top. Cover the super with a bee proof net or gauze, and replace the roof in an offset position. This will provide adequate protection and ventilation. Porous, bee proof material can be used to completely cover the hive holding the material down with stones or bricks; in this situation it is advisable to block the hive entrance. If hot weather is forecast place a frame feeder, containing syrup or water, or a wet sponge in the super. Do not cover the hive with any dark material that is likely to increase the heat gain in the colony.

If you find that spraying has taken place without you being informed, and you find large numbers of dead and dying bees outside your colonies, you should carry out the following actions:

1. Photograph the dead bees and the crop that you believe has been sprayed, including features in the photograph that will enable the location of the field to be confirmed during any subsequent investigation. Include the time and date on the image as is possible with digital and some other film cameras.

2. Find a witness who is willing and able to corroborate your experiences.

3. Take around 300 sample bees from each affected colony, place

each sample in a plastic bag and freeze them as soon as possible. Label the bags with the colony identification number, the date and time of collection and a description of the condition of the bees when you found them.

4.   Inform your local bee inspector or the National Bee Unit as soon as possible. They will advise you on the subsequent actions to be taken.

5.   Do not accept offers of compensation in return for not informing the authorities, and explain you are obliged to inform them in any event.

B - HYGIENE

THE SHOOK SWARM (FRAME CHANGE) TECHNIQUE AND ITS ROLE IN DISEASE CONTROL

This somewhat drastic technique involves the shaking of a colony of bees onto undrawn foundation and feeding them with (2 parts sugar : 1 part water) syrup to help them draw out comb. The principle is that the bees are separated from their combs in which there may be accumulations of spores, bacteria and other pathogens which can act as a source of infection. It should be carried out as early as possible in the season and not later than July. The technique is beneficial in the control of EFB, AFB, Nosema, Chalkbrood, Varroa and possibly a number of bee viruses. It ensures that there is a break in the rearing of brood allowing the beekeeper to use other control measures against diseases carried by the bees.

First prepare a brood box with new, undrawn foundation and place it next to the old colony. Remove each of the frames in the old colony and shake all the bees, including the queen, into the new box. Then place the new box on the site of the old one; any bees in the air will soon return to it. It is advisable to place a queen excluder above the floor and below the new brood box to prevent the queen from absconding.

If the technique is carried out correctly the bees are stimulated to produce good quality new combs. Preferably burn the brood frames you have removed in order to destroy the pathogens mentioned above and sterilize the old brood box to prevent the transmission of disease to the new combs. One disadvantage of the shook swarm technique is that if the queen is lost or damaged during manipulation because no brood is present the colony will be lost.

REGULAR BROOD COMB CHANGES

There is no doubt that the regular and systematic changing of brood combs is a key beekeeping practice that all beekeepers **must** adopt.

There are a number of diseases where the causative organism is voided onto or accumulates in the comb and constitutes an ongoing source of re-infection. Such infection potential can last many years. Examples include Nosema, EFB and AFB. Nosema is associated with a number of bee viruses and so any reduction in the infection potential for Nosema will also reduce the associated bee viruses.

The shook swarm technique can be used by beekeepers in their Integrated Bee Health Management plan to prevent the development of EFB; it can also be used to control Nosema. The technique requires the replacement of all of the combs with foundation and the bees have to draw out the foundation before the queen can recommence laying eggs which consumes both time and energy for the bees.

Unless there is an overriding reason to carry out the shook swarm technique it is easier for the beekeeper to institute a process of progressive brood comb replacement, for example changing three frames per year in each brood box. A good general rule of thumb is to hold the frame up to the light; it should be replaced if light cannot be seen through it. An alternative approach is to colour code the frames using the same colour marking scheme as is used for marking queens. The age of the combs can then be easily tracked.

Inserting single frames of foundation into a brood box containing drawn comb often results in misshapen combs being drawn from the foundation. Artificial swarming is a good way to get a brood box full of very straight comb except the initial comb with brood and queen used to hold the bees in the artificial swarm.

Access to borage or similar high nectar yielding crops stimulates bees to produce good straight combs. Brood boxes fitted with foundation and drawn can be used as honey supers and the combs then be either extracted, retained for use as winter stores, or used to replace old combs identified as needing replacement. Wax can be recovered from old brood combs however the amount of wax is relatively small. See Section 9 –Recovering and using Beeswax.

If the combs are no longer required they should be burnt or placed in domestic refuse bags and sealed to prevent bees gaining access and possibly picking up disease present on the combs. If the frames are to be reused they should be thoroughly cleaned and advice on this can be found in the following section on Cleaning Equipment.

CLEANING EQUIPMENT

Keeping beekeeping equipment clean is a vital part of preventing the spread of disease between colonies. Beekeepers who visit and work with other beekeepers should be especially vigilant to avoid spreading disease. Always insist that the visitor who comes to inspect your colonies for disease puts on new disposable gloves and sterilises their hive tools before allowing them access into your apiary. Hive tools and frame scrapers can be cleaned by soaking in a 0.5% sodium hypochlorite based product solution for 20 minutes.

Bee suits including hoods should be washed frequently, preferably with a non perfumed washing product. Usually bee hoods need to be washed separately because of their more delicate nature and remember that angry bees will deposit their pheromones and alarm substances on the hood and veil and if these are not completely removed in the washing process they will influence the behaviour of bees next time the hoods are used. Special attention should be paid to gloves; ideally they should be cleaned with a domestic bactericidal spray. Washable plastochrome and latex (household rubber gloves) are easy to keep clean and disposable surgical gloves are the most practical. Washing leather gloves is difficult and beekeepers should consider whether to use other materials to ensure disease control. Wellington boots and other footwear should also be washed in soapy water. It is amazing how much honey or sugar syrup ends up on a beekeeper's boots.

HONEY HOUSE CLEANLINESS

Not all beekeepers will have the luxury of a dedicated honey house but wherever honey is being extracted or processed it is essential that the beekeeper adopts the best hygiene practices possible.

Your honey house or other extracting areas should be vermin free, and if vermin are suspected suitable bait traps should be used (carefully following the label instructions on the products) and inspected and changed when required. Surfaces should be cleaned with proprietary products suitable for use in food preparation areas. Honey buckets, containers, sieves, cloths, extracting and bottling equipment should all be washed with products suitable for contact with food. Always wear clothing, including gloves and footwear, suitable for use in food handling areas.

For more detailed guidance contact your local beekeeping association, national bee health authority or local environmental health authority.

C – COLONY REPRODUCTION

PRODUCING AND USING NUCLEI

Nucleus colonies can play an essential role in effective bee health management. They contain frames of the same length and depth as those of full sized colonies and have fewer frames in each hive or colony. Typically they will contain 3-5 frames. This kind of nucleus should be distinguished from that of 'baby' or 'mating' nuclei which are very much smaller and are primarily used for queen mating.

Nuclei can be prepared from late spring to late summer using a healthy parent colony. Those prepared in the spring require the parent colony to be fed a 1:1 ratio of sugar to water (by weight) when daytime temperatures are consistently above 15°C. Protein supplements in the form of pollen patties can also be fed but are unnecessary if there is sufficient pollen forage available. Nuclei prepared later in the summer after the main nectar flow do not need the parent colonies to have been fed beforehand, if they are well provisioned they can be split with no preparation.

A typical 5 frame nucleus is composed of
- Two frames containing eggs
- One frame containing sealed brood
- One frame of pollen and honey
- One frame of foundation

A typical 4 frame nucleus is composed of
- One frame containing eggs
- One frame containing sealed brood
- 2 frames containing pollen and honey

Including the bees found on the frames in both cases

The queen remains in the parent colony and another queen is needed for the nucleus. This could be a queen purchased from a supplier, a queen raised by you from a queen cell, a virgin queen, a mated and laying queen, or a queen you want to keep for breeding purposes. Creating nucleus colonies is a good technique to use in swarm management as it alleviates the swarming tendency in crowded colonies.

The nucleus is basically a controlled swarm where you have moved bees from the parent colony thereby reducing colony congestion and the stimulation of the swarming impulse. The making of a 5 frame nucleus some 2-3 weeks before the swarming season begins greatly reduces the tendency for the colony to swarm.

A supply of nucleus colonies is a good way to keep productive colonies strong through periodically moving the bees and brood into the nuclei and uniting a nucleus to the production colony. It is essential to know the disease status of the colonies otherwise such actions could exacerbate the spread of disease.

Probably the best reason for having nuclei available is as a reservoir of queens that can be used to replace those in production colonies should they fail. Nuclei over- wintered are especially valuable although it is not easy to ensure their survival. Nuclei can also be used to augment weak colonies although it is essential to determine if the colony has become weak in the first place through disease, otherwise the nucleus will become diseased from contact with the weak stock.

Care is required in the management of nuclei. They should be protected from the possibility of robbing, e.g. plugging the entrance to the freshly prepared nucleus with dry grass for a few days. Nuclei expand very rapidly in good conditions and will soon outgrow a 5 frame nucleus box so always be prepared to move the colonies into full sized boxes. Expanding nuclei can quickly run out of stores unless there is a good nectar flow, so always check their food status.

HANDLING PULLED VIRGIN QUEENS

A pulled virgin queen is a young queen that has emerged from her cell after it has been removed (i.e. pulled) from the hive. Such cells can be utilised by putting each cell inside individual match boxes, kept warm (around 36°C) and allowing the queen to emerge inside the box. If a colony does not have an adult queen, a pulled virgin queen can be run into the colony from the matchbox if she has not met any other bees since emerging from her cell and she will almost always be accepted. Virgin queens can sometimes be seen emerging from their cells during an inspection at swarming time and they can be transferred to matchboxes and used in the same way. The matchboxes should be kept in a warm, dark place and the queens given a drop of diluted honey each day. Care should be taken opening the boxes and the easiest way to prevent the queen being lost is to open the matchbox tray against a piece of clear perspex through which she can be easily seen. Pulled virgins can be successfully introduced into bad-tempered colonies in which the queen could not be found.

UNITING QUEEN-RIGHT COLONIES

Before uniting colonies decide why you want to unite them. Colonies may be weak for a number of reasons including disease and it is bad practice to unite a diseased colony with a healthy one. The best time to combine colonies is during periods of abundant forage, but if feeding is necessary the colony can be fed with syrup using a frame feeder after the uniting process is complete.

Ideally the colonies to be united should be of equal strength, however, often the need is to unite a weak and a stronger colony. Always place the stronger colony below the weaker one. One of the colonies should be de-queened, otherwise the bees in the united colony will select the queen that is to survive, and if the queens fight the surviving queen may be injured. De-queen the colony immediately before it is moved.

There are a number of methods for uniting colonies; the most consistently successful and least stressful method involves the use of a sheet of newspaper placed between the two boxes containing the colonies to be united. The newspaper is perforated in several places to enable the bees to chew through it to open up a contact between the colonies. The bees then begin to mingle, the colony odours mix and uniting results. The presence on the alighting board of chewed pieces of newspaper is a good indicator that the uniting has been successful.

One practical problem with this technique is keeping the newspaper in place on top of the bottom brood box long enough to move the second box before the wind blows it away. The use of a queen excluder over the paper keeps it in place. Ensure that the queen is below the queen excluder before uniting. If the queen you wish to keep is in the upper brood box above the queen excluder she cannot move down with the bees during the uniting process and it will not succeed. A simpler method involves spraying the bees in both colonies with a sugar syrup 1 water: 1 sugar (by weight), and then combining the best frames from each colony into one brood box, shaking the bees on the remaining frames into the brood box containing the queen. A disadvantage of this is that even when one colony has been de-queened, the bees from the mixed frames may fight.

Do not unite a colony of egg-laying workers with another colony.

D –GENERAL TECHNIQUES

MOVING BEES

Moving colonies is an activity that, if handled badly, can cause bees much stress. It can provoke an outbreak of Nosema, and the potential for colony loss if the colony is exposed to high temperatures through lack of ventilation, space and water when travelling. If the colony temperature is raised to the point when wax comb loses its structure bees can drown in honey and nectar.

Evening is the best time to prepare and move colonies. The hive parts need to be fixed using strapping, spring clips, staples or wooden laths so the hive can be moved in one block. The use of staples or laths involves hammering on the sides of the boxes and this will disturb the bees and prompt them to investigate. It also involves holes being made in the hive woodwork. Strapping colonies with devices such as a SpanSet® is quicker and more convenient.

If you wish to move a colony within the same apiary select a frame of brood from the colony and place it into a brood box covered by a roof. Plug the entrance of the original hive with dry grass or a sponge, strap it up and move it to its new location. Place the new brood box on the original site to collect any late foragers. It can be left in place for a couple of days after which the brood box with any bees that have returned to the original colony site can be united with the original hive in its new location using the newspaper technique.

If the colonies are to be moved to a new location away from the apiary prepare them with a travelling screen on top of the top box and secure the hive and screen. Stow the cover board and the roof separately from the hive. Do not place them on top of the colonies as this will prevent adequate ventilation and risk heat stress. Then, preferably in the evening, block the hive entrances with a strip of foam rubber.

If the colony is to be moved in the daytime and the weather is hot, it may be necessary to stop periodically and spray clean water over the top frames to keep the colony cool. If it is raining cover the tops of the hives with a loosely draped but secured sheet of plastic to ensure good ventilation. This can be achieved by inserting bearers or hive parts, secured below the plastic sheet. Take extra care when driving to minimise the amount of disruption experienced by the bees during the move, especially if the colonies are being transported in a trailer without suspension.

On arrival at the new site place the hives where required before removing the entrance closure and releasing the bees. Before leaving the site count the number of entrance closures you have removed and ensure that they match the number of colonies, also check all the hives have their roofs in place and if necessary are weighted down, bricks or rocks are suitable, so take some with you if they are not available at the site.

CLEARING BEES FROM HONEY SUPERS

There are a number of techniques that can be used to clear bees from supers of honey. Typically they involve the insertion of a board fitted with one way bee escapes or valves; after the bee has passed through the valve it cannot return back into the super. The valves may be spring clips as in Porter Bee Escapes or cones. Ideally clearer boards should be dedicated to the clearing of bees and the valves kept clean of wax and propolis, and correctly adjusted. It takes around 24 hours to clear the bees; clearer boards can be inserted at any time of day and will clear faster when the ambient temperature is cool because the bees cluster more in the brood box.

Combs can be removed individually from supers and the bees shaken from them before they are placed in a bee proof box. Clearing can excite the bees and promote the onset of robbing. The process of clearing bees from a heavily populated super can be speeded up by controlled jarring of the super on the edges of an upturned roof on the ground. This will dislodge a significant number of bees into the top of the roof which can be transferred back into the colony. Use of the smoker can help to drive some of the bees into a lower part of the colony, however it can also result in the bees immediately entering the cells to ingest honey or, more importantly smoke particles can contaminate the comb and its honey. Several chemical repellents are available but they are smelly and sometimes messy to use.

Bees can be removed from supers set up on their sides and air blown through the frames from a portable leaf blowing machine. Unless you have a large number of supers to clear, this method is not very practical for most beekeepers.

THE USE OF DUMMY BOARDS
(also known as dummy frames)

Wood swells and shrinks in response to its moisture content and there can be situations where it is a tight squeeze to get the desired number of frames into the boxes. Often this results in bees being crushed or rolled when the frames are removed from the hive during inspection, as well as causing a mess if the face of the comb is abraded and the honey and nectar starts to run out of the comb. Dummy boards are full sized frames with 10 mm wide sides and top bars. They are fitted with a solid sheet of plywood instead of foundation or comb, and are inserted to replace one or more of the outer combs. They will prevent the outer faces of the outer combs from bulging, and stop the bees from producing brace comb with the side walls of the hive. When carrying out an inspection the dummy frames are removed to create space in which to slide and manipulate the combs more easily. After removing the dummy board the first frame is lifted out, if the queen is not present the bees can be shaken off the comb into the space created. Lay the frame aside and move the next frame along.

Dummy frames can be used for enclosing a brood nest of only a few frames in a full sized brood box to prevent heat loss and the build up of brace comb which can restrict ventilation, they can also improve insulation for the developing brood. Dummy boards can be made to fit in super frames.

VENTILATION OF THE COLONY

The ventilation of colonies is an important aspect of integrated bee health management.

Honey bee colonies are well adapted to maintaining cluster and brood nest temperature and humidity. Beekeepers should be aware of their requirements so that colonies can develop healthy brood, maintain wholesome food reserves and prevent conditions which might encourage the development of diseases and other debilitating conditions, e.g. fermenting honey stores and mouldy pollen.

Mesh floors

One factor that has delayed the adoption of mesh floors by beekeepers is the belief that such floors expose the colonies to extremes of cold and humidity. These are indeed factors which affect a colony but cold air falls and is lost through a mesh floor. Warm moist air, also associated with higher carbon dioxide concentrations, promotes yeast fermentation, mould spore germination and growth and also activates dormant spores of Chalkbrood that may be in the comb. The use of quilts or other insulation placed on top of the cover board will block ventilation and exacerbate the situation leading to increased moisture and carbon dioxide.

FEEDING WITH FUMIDIL®B TO CONTROL NOSEMA

Fumidil®B powder does not readily dissolve in water, and sugar syrup is needed to enable the Fumidil®B to be taken up by the bees. Mix the powder (see instruction leaflet to obtain correct mixing rate) with dry sugar, add warm water (<50°C) and make a sugar syrup of 2 parts sugar to 1 part water by weight and feed it to the bees in the usual way for winter feeding.

Alternatively apply a weak sugar solution of this mixture using a small hand spray. In this method mix one third of the 25g container of Fumidil®B (3 colony pack) or one dessertspoonful with 2 tablespoons of sugar and dissolve in 0.57 litres (one pint) of warm water (<50°C) add 3.4 litres (6 pints) of warm water. This quantity will treat 7 colonies. The treatment should be started about 3 weeks before the start of winter feeding. Spray the liquid over the frame tops in situ, over the bees and on the exposed parts of the four walls of the hive so that it runs down them. Repeat the operation after 6-7 days and again after another 6-7 days making up fresh solution each time.

For a small number of colonies such as three, mix one teaspoon of Fumidil®B with a tablespoon of sugar and dissolve it in 0.28 litres (0.5 pints) of warm water then add 1.4 litres (2.5 pints) of warm water.

If faecal spotting is noticed during the winter carry out emergency treatment using the spray method during mild winter spells when the bees are flying to defecate. Use 0.28 litres (0.5 pints) of liquid per hive and do not spray the walls of the hive. A full treatment of the colony can be made in the spring when the cluster breaks up.

RECOVERING AND USING BEESWAX

The practice of regular brood frame changing is essential as part of a bee health management strategy. It will result in an increased amount of old comb that will yield low amounts of useable wax, whichever extraction method you use. Also there is always the possibility of reintroducing the spores of brood diseases into new foundation made from recovered wax, as well as residues from any treatments applied to the bees during the life of the frame.

The removal of the old brood combs from the frames can be messy and any spores on brood combs and frames will be spread during the handling of the combs. The most effective way to deal with old brood combs and wooden frames is to burn them, preferably in the open air. This will stop the transmission of disease present on the combs. The ash, when cool, can be put in a sealed bag and placed in the domestic waste for collection and disposal by the local authority. From the economic and environmental perspectives the discarding of the combs and frames is better than removing the combs and washing and sterilising them because it is time-consuming and creates contaminated water for disposal.

The residue of processing your honey by melting it from comb in the supers leaves a lot of sticky wax. It is worth taking the time and effort to extract, wash and drain the wax until it is dry enough to be stored for the next processing stage. Always use clean rain water if possible as water from the tap may contain calcium which will bind with some of the constituents of the wax and affect its natural properties. This wax is less likely to contain any residues and only a small chance of any brood disease spores.

During days of warm sunshine it is possible to melt reclaimed wax in a solar wax extractor allowing it to drain into a collecting container where it will cool into a block over night. The disadvantage of this technique is that the temperature cannot be controlled. Alternatively the wax can be processed using steam stripping, collected in containers and allowed to cool naturally. This process produces a liquid honey and water mix on which the block of wax floats. The block is removed, drained and the detritus left clinging to the bottom, known as slum gum, is removed; after drying the wax block is ready to be stored. It can be further processed into foundation, candles, polish and other bee products, exchanged at a number of bee equipment suppliers or at shows for new foundation, beekeeping equipment or money.

SOME CONSIDERATIONS WHEN BUYING BEEHIVES

TABLE 20 Comparison of the dimensions of different types of hives

Type	Brood body dimensions	No. of frames in the brood body	Comb area in brood body	Approx. no. of worker cells in brood body	Top or bottom bee space	Notes
WBC	460 x 419mm	10 BS	2000sq in	45,000	Bottom	Double walled
Smith	464 x 416 mm	11 BS with short lugs	2200sq in	50,000	Top	Popular in Scotland
National and Modified National	460 x 460mm	11 BS	2200sq in	50,000	Bottom	Most popular hive in the British Isles
Langstroth	508 x 414mm	10 self spacing	2750sq in	62,000	Top	The world's most used hive
Modified Commercial	464 x 464mm	12 self spacing, normally 11 + a dummy board)	3300sq in (with 12 frames)	75,000	Bottom	Hive parts interchangeable with national except for the frames
Modified Dadant	508 x 470mm	11 self spacing	3750sq in	85,000	Top	The largest hive in common use

Imperial equivalents

413mm	16 ¼ in
419mm	16 ½ in
425mm	16 ¾ in
462mm	18 ⅛ in
464mm	18 ¼ in
470mm	18 ½ in
508mm	20 in

Top and Bottom bee space configurations

To ensure that frames can easily be removed from boxes during manipulations it is essential that the bee space is maintained between all the separable parts, in particular between the tops of frames in the lower box, and the bottom of frames of the box above. Hives are either top or bottom bee spaced. In bottom bee space hives the top bars of the frames are flush with the top of the box and the bottom edge of each box must extend 6mm below the bottom bars of the frames it holds. In top bee space hives the top bars fit 6 mm down from the top edge of the box and the bottoms of the frames above lie level with the bottom of the box. Some hive patterns, e.g. the Modified National can be supplied with a top or a bottom bee space configuration. This aspect should be carefully examined if you are considering buying second hand equipment and once you have chosen the type of hive to suit you make sure all the equipment you buy is compatible with it.

TABLE 21 The advantages & disadvantages of different types of hives

Type	Pros	Cons
WBC	Double walled (good for protection)	Small brood chamber, suitable only for colonies no larger than 45,000. Takes longer to manipulate and move from site to site. Requires maintenance if painted.
Smith	Simple single walled structure.	Hand holds difficult.
National / Modified National	Single walled. Good hand holds.	Suitable only for colonies <55,000, unless 2 brood boxes used; or use a 14 x 12 inch brood frame. Complicated structure.
Modified Commercial	Can support large colonies.	Hand holds difficult.
Langstroth	Can support larger colonies.	Hand holds difficult.
Modified Dadant	Can support very large colonies.	Hand holds difficult. Very heavy to lift and move.

EXTENDING THE LIFE OF THE HIVE

Hive bodies are expensive, careful use and regular maintenance will be rewarded by the reduced need for replacement. The following comments apply to hive parts made of wood which is still the predominant material used.

Wood is a superb material in terms of its strength to weight properties but it is vulnerable to biological attack, in particular fungal degradation. Insect attack only usually occurs in wooden hive parts which have been stored and unused for a long period of time. Fungal degradation arises if the moisture content of the wood exceeds 22%. Wood is hygroscopic and if placed in wet conditions will absorb moisture to a level well above 22%. Dampness in colonies is often the cause of mould and other conditions, so keeping the wood as dry as possible is beneficial.

As with other biocidal products wood preservatives require an authorisation to enable the manufacturers to be able to place them on the market. In the UK the wood preservative product is given an HSE Number and this can be seen on the product label which also defines where and how the product can be used. Check it to see if it states that the product should **not** be used on beehives or beekeeping equipment.

The brush application of wood preservatives to the exterior of the hive boxes will help extend the life of the boxes but it is better to buy pressure treated wood for hive stands and use them to keep the hive out of contact with the ground. Ideally the floor height should be about 0.7 m (28 inches) above the ground. Wood stains and other types of coatings can be used however alkyd gloss paint systems can trap moisture below the coating leading to blistering.

KEEPING RECORDS

Colony development

The health of your bees depends on regular colony monitoring and to ensure a full record of the colony status, health, disease treatment and control it is important to make careful notes after each inspection. The way in which you keep records is a personal choice, ranging from an index card system to a computer spreadsheet.

Each numbered colony should have a separate entry in which some basic information is recorded; this should include:
- The colony number and apiary location.
- The origin and history of the queen / swarm / nucleus.
- The colour of the queen marker.
- The age of the queen.

At each full inspection the following information should be noted
- Weather conditions.
- Presence (or evidence of) or absence of the queen.
- If the queen is marked (which colour) and if her wings are clipped.

- The number of frames with all types of brood and the stage they have reached e.g. eggs, sealed larvae.
- An assessment of available stores.
- The condition of the adult and larvae with respect to health and disease.
- The temper of the colony and the behaviour of the bees on the comb.
- Amount of honey removed.
- Amount of syrup fed.
- Type of treatment given and when the treatment is finished (see section on treatment records).
- Data on Varroa mite numbers (from drone uncapping technique or natural mite mortality drop).
- Any other points of interest.

Many beekeepers develop a form of shorthand to record their observations and Donald Sims in his book 'Sixty Years with Bees' contains some useful advice on symbols to use when recording colony inspections and events.

Treatment records

Beekeepers are required by law to keep records of all substances they use to treat honey bee colonies for disease and conditions control. The objective is to be able to trace any honey bee products from a colony that has been treated with a particular substance or product if a concern is raised.

Records should include the following information for each colony.

- The name of product or substance used.
- The manufacturer's name.
- The Lot or Batch Number of the product used.
- The expiry date for the product.
- The rate of application of the product.
- The start and end time of the treatment period.
- The manner of disposal of the used treatments and associated packaging.

DEALING WITH BEE STINGS

There is always a risk of being stung during manipulations and there are a few simple rules to reduce this risk.

- Wear a clean bee suit and clean gloves to ensure there is no bee venom or alarm pheromones present from previous beekeeping activities.
- Avoid carrying out manipulations on cold, wet, thundery or windy days.
- Avoid being hot, sweaty and smelly.
- When handling bees do not use any perfumed product, even hair shampoo will be noticed by the bees.
- Avoid drinking alcohol before handling bees, the smell of it disturbs them.
- Handle the combs of bees with firm, gentle movements, avoiding sudden actions.
- Keep bees that are docile and easy to handle.

Dealing with bee stings is an important skill for beekeepers. They need especially to recognise when they are becoming sensitized to the bee venom and require medical treatment because of the risk of suffering an anaphylactic shock.

Treatment of bee stings-dealing with the local skin reaction

When the sting penetrates the skin and injects venom the cells closest to the point of entry are killed. Remove the sting by scraping it from the skin either with a hive tool or thumbnail. Do not pinch or squeeze the sting to remove it as this will force more poison from the venom sac into the wound. Take a double dose of oral antihistamine immediately after being stung, e.g. 2 x 4mg tablets of Piriton®. Use an ice pack or a pack of frozen peas to relieve the pain and help reduce the swelling. In addition take a single dose of painkiller e.g. paracetamol. **Please note this information is given in good faith and you should ensure that you can take such measures without prejudicing your health.**

Contrary to general belief beekeepers can become sensitised to bee stings and develop Type 1 hypersensitivity leading to an anaphylactic reaction. Practice your beekeeping to reduce the possibility of being stung and be alert to the early symptoms after a sting. Consult your doctor for referral to an allergist if there is a change in the symptoms you experience.

Helping third parties who are stung by honey bees

If a person is stung and is distressed act as follows:

1.   Move the person away from the hives.
2.   Scrape out the sting as quickly as possible.
3.   Sit the conscious person upright, loosen their clothing at the waist and neck and encourage them to remain calm.
4.   If there are signs of difficultly in breathing, light-headedness or a general reaction to the sting ask the person if this is normal and if they have any medication provided by their GP (e.g. antihistamine tablets). If so let them medicate themselves.
5.   Stay with them and encourage them to breathe in and out regularly.
6.   Phone for an ambulance giving location details and stating the emergency concerns a **bee sting reaction.**

If the person is **unconscious** carry out the following actions:

1.   Loosen tight clothing and place them in the recovery position, tilting the head back slightly to obtain a good airway.
2.   Check they are breathing and you can feel a pulse in the side of the neck.
3.   Do not try to give the person any food or drink.
4.   Call an ambulance as above stating that the person is **unconscious following a bee sting reaction.**

USING THE SMOKER

A clean and efficient smoker is a prerequisite. It is well worth spending a little extra to purchase a smoker which has the firebox enclosed in a wire cage. This will protect you from burns to your hands and knees. You can hold the lit smoker between your knees while you carry out your manipulations.

The fuels used to generate the smoke should not be harmful to bees or to beekeepers. It is surprising how easy it is to inhale the smoke during beekeeping manipulations. Proprietary products made from impregnated cardboard rolls, straw pellets, hessian sacking and cotton cartridges can be safely used. With other materials take care that the smoke does not have injurious properties. Use windproof matches or outdoor lighters to help get the smoker fuel ignited quickly. Smoke should never be used in an attempt to subjugate the bees, as they could react violently and attack the beekeeper, bystanders and animals.

Always take care when emptying your smoker especially if it still contains glowing embers when you discard the ashes onto the ground. In dry windy conditions they can be fanned into flames and start a fire. This is an especial hazard on heather moorland. A plug of damp grass inserted into the nozzle will starve the firebox of oxygen and the fire will go out. Discard the ashes onto the ground and scuff them with your foot and bury them in the soil. If any of your hive roofs are made of bitumen felt be aware that the base of the smoker might melt the felt.